Logics of War

A volume in the series

CORNELL STUDIES IN SECURITY AFFAIRS

edited by Robert J. Art, Robert Jervis, and Stephen M. Walt

A list of titles in the series is available at www.cornellpress.cornell.edu.

Logics of War

EXPLANATIONS FOR LIMITED AND UNLIMITED CONFLICTS

ALEX WEISIGER

Cornell University Press

ITHACA AND LONDON

Cornell University Press gratefully acknowledges receipt of a subvention from the School of Arts and Sciences at the University of Pennsylvania which aided in the publication of this book.

First published 2013 by Cornell University Press

Printed in the United States of America

Library of Congress Cataloging-in-Publication Data

Weisiger, Alex, 1977–
Logics of war: explanations for limited and unlimited conflicts / Alex Weisiger.
        p. cm. — (Cornell studies in security affairs)
    Includes bibliographical references and index.
    ISBN 978-0-8014-5186-7 (cloth : alk. paper)
    1. War—Causes.   2. Limited war.   3. Low-intensity conflicts (Military science)
4. Total war.   I. Title.
    JZ6385.W45   2013
    355.02—dc23   2012043954

Cornell University Press strives to use environmentally responsible suppliers and materials to the fullest extent possible in the publishing of its books. Such materials include vegetable-based, low-VOC inks and acid-free papers that are recycled, totally chlorine-free, or partly composed of nonwood fibers. For further information, visit our website at www.cornellpress.cornell.edu.

Cloth printing   10   9   8   7   6   5   4   3   2   1

Contents

Acknowledgments

This book has been in the making for many years, and in that time I have accumulated an unusually large number of debts. Columbia University was and is a wonderful place to study international politics—this book benefited tremendously from the mentors and friends from whom I learned there. Robert Jervis is as brilliant an adviser as one could want—I hope that anyone who reads this manuscript will be able to discern the ways in which his ideas and suggestions have influenced me. I benefited equally from an excellent cohort of then-junior faculty—Page Fortna, Erik Gartzke, and Tanisha Fazal—whose comments and suggestions on matters both large and small consistently improved the argument and evidence that I present. Perhaps even more important was the community of fellow graduate students, most notably the regular attendees at the international relations seminar, especially Marko Duranovic, Leila Kazemi, Thania Sanchez, Ivan Savic, Zach Shirkey, Jessica Stanton, Matt Winters, Catharina Wrede-Braden, and Maria Zaitseva, whose early dissatisfactions led me in directions that greatly improved the manuscript. I also benefited from regular conversations with Josh Baron, Megan Gilroy, Georgia Kernell, Paul MacDonald, Joe Parent, Sharon Sprayregen, and Rob Trager. Beyond Columbia's walls, I was fortunate to have enjoyed the financial and intellectual support of the Belfer Center for Science and International Affairs and the Olin Institute at Harvard University, who gave me two years to complete my research and then begin revising it into this book in a particularly supportive academic environment.

If I was fortunate in the opportunities I had in graduate school, I have been equally fortunate to find a similar intellectual home in the political science department at the University of Pennsylvania. Ed Mansfield and Avery Goldstein have been everything that one could want in mentors, welcoming, generous with their time, and cogent in their advice. I am particularly indebted to Ed and to the Christopher H. Browne Center for funding a book conference, and to Hein Goemans and Dan Reiter (along with other participants in that conference) for reading the full manuscript and providing extensive, detailed, and constructive advice. I bear a similar debt to Robert Art and an anonymous reviewer, whose suggestions improved the argument and evidence advanced here, and to Roger Haydon, Karen Laun, and others at Cornell University Press for comments that have greatly improved the presentation. Among my colleagues, I have benefited from conversations with Daniel Gillion, Mike Horowitz, Ellen Kennedy, Matt Levendusky, Ian Lustick, Marc Meredith, and (once again) Jessica Stanton. Rosella Cappella, Barbara Elias, Krishan Malhotra, and Bashak Taraktas all provided invaluable research assistance. Last, but certainly not least, Sarah Weisiger has provided constant support and repeated (and necessary) reality checks.

It has been a long time since I started working on this book, but its genesis in many ways lies even earlier. My undergraduate adviser, Stephen Krasner, played a critical role in my transition from student to researcher. Well before him, my parents, Richard Weisiger and Jane Martin, shaped the way in which I think about the world. Their willingness to endure unending questions and their encouragement to seek out answers for myself, more than anything else, led me to a life that I would not trade for any other. It is to them that this book is dedicated.

Introduction

In the summer of 1866, Prussia and Austria went to war to determine who would be dominant in Germany; one battle, on July 3 at Könniggrätz, was sufficient to resolve the issue, and the war ended less than two months after it started. In the summer of 1914, Austria again went to war, this time to establish dominance over the South Slavs; this conflict lasted more than four years, with deaths in the millions rather than the thousands. The contrast between these two conflicts—wars that appeared quite similar at the outset but differed dramatically in their eventual duration and destructiveness—is hardly unique. To cite another example, in 1929, the Soviet Union invaded China, which was riven by internal conflict, to secure its hold on the Russian Manchurian Railway; this conflict ended quickly and at low cost. In contrast, Saddam Hussein's effort in 1980 to take advantage of Iran's temporary weakness following the revolution in 1979 to advance territorial claims along the Iran-Iraq border dragged his country into what turned out to be an eight-year war in which hundreds of thousands of people died.

With the benefit of hindsight, the relative severity and length of these wars may seem intuitive, yet for contemporaries there were good reasons to expect quite different outcomes. The outcome that occurred in 1914—expansion of a localized dispute into a general European war—was possible in 1866: Bismarck's policies were heavily influenced by the possibility that France or Russia might intervene on the side of Austria.[1] On the other hand, the conventional view of World War I has been that the initial participants expected the war to end quickly and decisively; from this perspective, the trench warfare that developed was a complete surprise.[2] In the Sino-Soviet case, Stalin apparently considered using the conflict over the Manchurian railway as a pretext to launch a broader war with the aim of replacing the Nationalist government in China with a communist one; Iran in fact (unsuccessfully) used the conflict with Iraq as an opportunity to spread the revolution.[3] Thus wars that appear quite similar at the outset

[1]

Duration

	Short	Long
High	Russo-Finnish War Persian Gulf War 29% of wars 3% of deaths	World Wars Iran-Iraq War 10% of wars 86% of deaths
Low	Falklands War Kosovo War 43% of wars 2% of deaths	Vietnam War Franco-Turkish War 18% of wars 9% of deaths

Intensity

FIGURE 1.1 The frequency and destructiveness of interstate wars

ultimately diverged dramatically, with some ending quickly and at low cost while others dragged on for year after painful year.

Indeed, the extremely destructive conflicts like World War I or the Iran-Iraq War are remarkably unrepresentative, as figure I.1 demonstrates. This figure differentiates between long and short wars, where long wars last over a year, and between low- and high-intensity conflicts, where intensity references the *rate* at which deaths accumulate.[4] Most wars are short, lasting less than a year; indeed, median war duration is about four months. Very few are both long and intense, as was the case in the World Wars and the Iran-Iraq conflict. Yet these few wars are responsible for a disproportionate amount of human suffering: the 10 percent of wars that are both long and intense are responsible for 86 percent of all deaths in battle over the past two centuries.

What separates the few unusually destructive wars from the many that are limited, either in duration or in intensity? The primary goal of this book is to answer this question.

THE ARGUMENT WRIT SHORT

My answer to this question relies on the observation that there are multiple logics of war. In the jargon of academia, war is characterized by equifinality—there exist multiple independent causal pathways that can lead to war.[5] We can think of all of these paths as containing individual causes of war, with each cause being a reason why the adversaries in the conflict would rather fight than accept peace on the opponent's terms.[6]

[2]

Each of these causes thus is individually sufficient to bring about fighting, although multiple causes may be at work in any real-world case. When we move from the question of why war begins to examining variation in its destructiveness, however, not all causes of war are equal. Some are relatively easily dealt with, as leaders or publics come to realize that the war that they expected is not the one that they are going to get. Others push leaders toward increased war aims and a higher tolerance for the costs of war, with the result that it may take many battles and much suffering before settlement can be achieved. In rare cases, one side in a war takes settlement entirely off the table, guaranteeing that war will continue until one side subjugates the other militarily.

Thus I argue that unraveling the logic of different causes of war allows us to explain the variation in war duration and severity. In developing this argument, I work from the bargaining model of war, which I discuss in greater detail in the next chapter. In particular, I examine three causal mechanisms that are believed to be particularly important in bringing about violent conflict: divergent expectations and mutual overoptimism, principal-agent problems in domestic politics, and commitment problems that generate an inability to trust one's opponent to live up to a political agreement. None of these causes are new—it would be highly disconcerting if in the more than two millennia that war has been studied major causes had eluded our attention. That said, our understanding of the implications of these causes of war for war's duration and deadliness remains imperfect. As it turns out, each is associated with a unique path from peace to war and back to peace, with the result that the wars that they bring about look dramatically different.

In particular, I argue, and find, that commitment problems produce unusually long and deadly wars. Most explanations for commitment problems focus on shifting power: a country that anticipates relative decline in the future must fear that its opponent's rise will eventually allow that opponent to impose painful political concessions, as once the opponent's rise is complete the now-weaker declining power will be more reluctant to fight. If the rising opponent is unable to allay those fears—a difficult proposition given the incentive for even hostile rising powers to claim that their intentions are benign while the rise is occurring—then war to prevent the rise from occurring may be an attractive option. The problem, however, is that if fear of decline is a cause of war, then war will continue until that fear is addressed, which will typically require either that the decline be prevented or that it come into being. Short of preventing the decline from occurring, a militarily undefeated declining power will typically prefer to continue to fight rather than settle. Moreover, preventing a decline from occurring will typically require an unusually large victory, meaning that the declining power's aims in these wars will be unusually high. As a result, these wars can endure for a long time, even when fighting is quite intense.

Moreover, I identify a different type of commitment problem, which I refer to as the "dispositional commitment problem," in which the belief that no agreement is credible exists not because of shifting power (a situational commitment problem) but because leaders conclude that their opponent simply is not deterred by the costs of war. An opponent that is by nature aggressive presents tremendous problems for war termination, however, as she can be expected to break any agreement as soon as it is convenient for her to do so. Thus leaders who sincerely believe that war is a consequence of an opponent's character will resist settlement on any terms that do not permit the replacement of the opposing leadership and, frequently, the remaking of the opposing society. This dynamic is thus associated with demands for unconditional surrender, which, although rare, are a feature of some of the worst wars in history.

I argue that the dispositional commitment problem—and hence a categorical refusal to negotiate with the existing regime—arises in a consistent fashion across cases. First, leaders who fear relative decline launch a conventional preventive war out of a belief that the target of their attack would otherwise attack them or impose unpalatable concessions in the future. In these cases, the belief in the opponent's hostile intentions, while often understandable, turns out to be inaccurate. Attacked by an aggressive opponent who claims that her aggression is justified by intentions that do not in fact exist, the target of the attack concludes that the stated justifications are simply rhetorical cover for a preference for naked aggression. Thus what starts as an aggressive war—given the initiator's preventive motivations—becomes a war to the death.

That commitment problems, either situational or dispositional, have the potential to produce unusually destructive wars is the central argument in this book. That said, a convincing account of the variation in interstate war destructiveness must also explain why noncommitment problem conflicts tend to be limited. I thus examine two additional mechanisms, which I argue are the most salient alternative logics of war. Both provide theoretically coherent explanations for costly conflict, but I argue that the wars that they produce inevitably will be limited in either duration or intensity. In the informational mechanism, fighting arises because the two sides have differing expectations about how the war will go: for example, because they disagree about relative strength or resolve, and hence are unwilling to agree to each other's demands. Once war begins, however, these expectations are put to the test, and at least one side inevitably must be disappointed by events on the battlefield or in the diplomatic arena. This disappointment forces leaders to revise expectations, and hence demands, bringing them closer until settlement is

[4]

reached. Fighting thus inevitably leads toward settlement, more quickly when fighting is intense.

Wars driven by domestic politics—in which leaders resort to war to pursue policies that the public would not have chosen in their stead—are similarly limited. The most prominent example of this sort of behavior is diversionary war, in which leaders attempt to use external conflict to distract attention from internal problems and thus improve the leader's hold on power, but this logic also covers situations in which leaders pursue policies that represent a definite minority interest in society more generally. While leaders can start wars that are not in the general interest, their ability to continue them depends on whether they can hold onto power and avoid being forced into a settlement that would, after all, be in the interests of the leader's constituents. A leader's ability to sustain this kind of war thus depends critically on her informational advantage over her citizens: if she can claim convincingly that continued war is in the general interest, then fighting will be unlikely to end. As in the informational mechanism, however, the revelation of information as the war continues limits the degree to which the leader can plausibly claim that further fighting is in the general interest; beyond a certain point, orders to continue fighting will be ignored or overridden, or the leader will be replaced by someone more willing to negotiate. Once again, this process will happen more quickly when fighting is more salient, implying that wars driven by this mechanism can be either long or intense, but not both.

The Contributions of This Study, or Why Another Book About War?

Given the number of books that have been written about war, what can another one add to what we already know? I argue that this book makes a number of significant contributions to our understanding of war. Most important, it addresses a question for which the field does not yet have an established answer. Political scientists and historians have long been interested in determining the causes of big wars: the events of July 1914 or the Munich Crisis in 1938 have been minutely dissected and reinterpreted. They have shown by contrast less interest in relatively minor wars, given their more limited consequences. Yet by focusing primarily on the big wars, they have, ironically, limited their ability to determine what it is that makes those wars so destructive. Similarly, because it focuses more on the onset of war than its conduct or termination, existing work is poorly positioned to explain why it is that particularly deadly wars were so hard to resolve.

[5]

This point becomes clear if we think about how existing work would lead us to answer this question. Clausewitz notes that left to its devices war tends to the extreme, but he provides little insight into why some wars become more extreme than others.[7] The Realist focus on systemic wars leads to a standard assumption that great power wars are unusually destructive, while non–great power wars are not.[8] To a certain extent this observation is true, of course, simply because the great powers have tended to have larger populations and thus be able to inflict and suffer greater numbers of deaths. Yet not every major power war is particularly destructive: France and Austria fought a war in the nineteenth century that lasted less than three months and brought about "only" twenty thousand deaths, while the Soviet Union and Japan fought two border wars in 1938 and 1939 that have been all but forgotten by history.[9] Moreover, some non–great power wars have been horrific: Paraguay most likely lost over half its population in a six-year war against Argentina and Brazil in the 1860s, and Iran and Iraq fought an incredibly destructive war in the 1980s.[10]

More recently, a statistical literature has emerged that directly contrasts wars of differing duration and severity, but it has done so at the expense of compelling theory.[11] Thus, for example, Bennett and Stam find that realist variables have the greatest influence over interstate war duration, while domestic political variables play a secondary, but still important, role.[12] With respect to war size, Cederman builds on an earlier finding that deaths in war follow a power law distribution to argue that war size is essentially random.[13] While empirically quite interesting, both these studies, and several others like them, do not address the central question of why leaders sometimes decide to settle wars and sometimes decide to continue fighting: for Bennett and Stam, the central question is how long it takes one side to triumph militarily, while for Cederman there seem to be no political decisions whatsoever.[14]

Yet the question of why some wars end quickly while others drag on for years is extraordinarily important. A tremendous amount of work has been put into understanding how policymakers might prevent wars from occurring or promote peace in ongoing conflicts, yet without an understanding of why it is that wars differ in their duration and severity we are limited in our ability to develop definitive recommendations for how to promote peace most effectively. This point is especially important given the observation that war is characterized by equifinality: a strategy that works quite well for dealing with one cause of war may be ineffective or even counterproductive for another. Similarly, a better understanding of the implications of different causes of war for war duration and severity may allow for better advice when it comes to the use of force. If we know, for example, that a given logic of war tends to produce

[6]

conflicts that are particularly bloody, we may be more willing to seek peaceful alternatives when that logic drives us to consider going to war.

At a far more prosaic, if still important, level, this study helps to fill a gap in the scholarly study of war more generally. In the past fifteen years, theorists have developed an impressive structure of theoretical work based on the bargaining model of war; it is largely because of the attractiveness of this theoretical logic that this model has become so prevalent in the academic study of violent conflict. That said, empirical applications and tests of bargaining model predictions have not kept up with the theoretical advancements.[15] As a result, we have not resolved disagreements among different theorists, as with the question—discussed in the next chapter—of how long a war driven by divergent expectations should last. Moreover, an increase in the range of studies that test key claims from the bargaining model empirically will generate greater confidence that the entire theoretical framework is useful. From this perspective, then, positive findings in this study can help to increase confidence in a whole range of findings that are similarly grounded in the bargaining model of war.

Clarifying Terminology

There is a significant potential for terminological confusion in this project, so it makes sense to define some of the key terms. With respect to theory, I discuss four different causal mechanisms, all of which bring together a number of related concepts. The situational commitment problem, which I also refer to as the preventive war mechanism, involves declining powers starting wars to forestall the undesirable implications of decline. The dispositional commitment problem, also referred to as the unconditional surrender mechanism, involves actors refusing to negotiate because they believe their opponent is by nature predisposed to aggression. In the informational mechanism, overoptimistic actors fight because of their divergent expectations about how the war will go. Finally, in the principal-agent mechanism, misbehaving leaders make use of their advantages in domestic politics to pursue wars that serve their interests rather than those of their constituents; this mechanism includes diversionary wars, but also other conflicts (which I refer to as "policy wars") in which the leader pursues an interest other than improving her hold on power.

On the empirical side, I use the term *destructiveness* to refer to all forms of suffering and devastation imposed by war. The most obvious indicator of destructiveness is total deaths (which I treat as synonymous to war severity), but I also examine war duration and the economic cost of fighting.[16] The argument that certain mechanisms are logically limited

[7]

focuses on both duration and intensity, where intense wars are characterized by frequent significant military clashes, or in other words by higher death totals *per unit of time*. This concept is thus distinct from war severity (deaths for the war as a whole): a war could be intense but not particularly severe, as for example in the Arab-Israeli Six Day War, if the period of fighting is relatively brief. Finally, I distinguish between limited and unlimited wars, where wars can be limited either because they are fought at a relatively low level of intensity (e.g. the Vietnam War) or because they are relatively short, while unlimited wars are both intense and long.

CHAPTER OUTLINE

The remainder of this book consists of three broad sections: theoretical arguments about the determinants of war duration and severity, quantitative statistical tests of hypotheses drawn from those arguments, and case study tests of both those hypotheses and further inferences that are not amenable to statistical tests. Chapter 1 presents the theoretical argument, starting with a summary of the bargaining model of war and then advancing my explanations for both unlimited and limited wars. I begin with preventive wars driven by the situational commitment problem, explaining why these wars will tend to be difficult to resolve once they have begun. I then turn to the dispositional commitment problem, which leads to sincere demands for unconditional surrender. I first explain why this behavior is a logical consequence of the belief that one's opponent is dispositionally aggressive—in short, that the opponent enjoys war—and then provide a novel explanation for this belief, grounded in the way that the targets of preventive war interpret their opponents' attacks. The final section of this chapter examines the other two mechanisms, in each case explaining both the logic of the decision to fight and the reason why wars driven by the mechanism will be limited in duration, intensity, or both. In addition to advancing general predictions about whether wars driven by the different mechanisms will be limited, I also derive specific hypotheses for each mechanism that allow for more convincing empirical tests.

Chapter 2 presents the quantitative analysis. This chapter begins with a discussion of the appropriate research strategy for testing the hypotheses developed in chapter 1, focusing in particular on the decision to combine quantitative and qualitative analysis. I then introduce the dataset used in the quantitative analysis, which combines standard international relations data with information collected specifically for this project. The main section of this chapter presents the statistical tests. I start by examining variation in war destructiveness, measured in three different ways (duration, battle deaths, and total spending), and then turn to more specific tests that examine the speed of settlement or conquest, the choice of military strategy, and the nature of war termination. The most important finding is that larger prewar shifts in relative capabilities, which I argue are proxies for anticipated future shifts, are associated with unusually destructive wars, in particular because these wars are unusually difficult to settle. Other results suggest that noncommitment problem wars will tend to be either intense or long, but not both.

The remaining chapters present my case studies. Chapter 3 examines the understudied Paraguayan War of the nineteenth century, in which Paraguay—a buffer state between regional powers Brazil and Argentina—launched an aggressive war against both of its neighbors that ultimately killed over half the Paraguayan population and that has puzzled historians ever since. I find that this aggressive and risky policy followed from a fear of decline created by its neighbors' economic and military rise and by their incipient alliance. The case is particularly useful for the analysis of dispositional commitment problems, as Brazil, but not Argentina, refused to consider negotiation with Paraguay; consistent with my predictions, the historical record demonstrates that Paraguayan fears of Argentina were well founded, but that those of Brazil were, if not unfounded, in fact inaccurate.

The next chapter examines World War II in Europe. The first section focuses on the sources of German expansion, which I argue arose from the belief, grounded in Nazi ideology, that Germany faced irreversible decline absent the acquisition of most of Eastern Europe. Moreover, by the late 1930s, Germany's rearmament and Stalin's purge of the Red Army officer corps created a situation in which Germany would never have a better opportunity to address Hitler's fears. Consistent with the commitment problem argument, Hitler had expansive war aims that he pursued through risky strategies and refused to abandon even in the face of military defeats. The second half of the chapter analyzes the Allied refusal to negotiate with Germany once the war was underway, focusing on the British decision not to negotiate after the fall of France in the summer of 1940 and the Allied decision to demand Germany's unconditional surrender. Again, the dispositional commitment problem provides a compelling explanation, as the Allies, who obviously did not share Hitler's ideological theories of international politics, did not understand the threat that he believed Germany faced and certainly did not intend to do what he expected. Given this disconnect, they concluded that the Germans were fundamentally aggressive, and hence that only a thoroughgoing reform of the German social and political system after Germany's unconditional surrender would produce sustained peace.

[9]

Chapter 5 presents shorter case studies of additional major wars, including the Crimean War, the Pacific War in World War II, and the Iran-Iraq War. These cases, although presented in substantially less detail than the Paraguayan and European World War II cases, provide an additional opportunity to see the commitment problem arguments in action. In the Crimean case, the British had strong preventive motivations for war that arose out of the fear that Russia was on the verge of acquiring Constantinople; their aggressive war aims and reluctance to settle followed from this fear. The Russians, however, understood the British concerns and thus remained open to negotiation, exactly as I would predict. In the Pacific War, the Japanese concluded that significant expansion was necessary to forestall decline; the Americans, who failed to understand this fear, responded to the Pearl Harbor attack with the conclusion that negotiation with Japan was futile. Finally, in the Iran-Iraq War, Saddam Hussein attempted to take advantage of a temporary window of opportunity associated with the Iranian Revolution; the Iranian response was to launch an ideological crusade designed to remake the Iraqi state and thereby eliminate the dispositional threat that they associated with Saddam. All three cases thus provide further support for the arguments about situational and dispositional commitment problems.

Chapters 6 and 7 address the concern that a convincing explanation for large wars should also be able to explain small wars by examining four relatively small wars, which turn out to be driven by the informational and principal-agent mechanisms. Chapter 6 contains case studies of the 1991 Persian Gulf War and the 1856–57 Anglo-Persian War, both of which match the expectations of the informational mechanism. In both cases, war occurred because the participants disagreed about relative strength or resolve, and in both cases the recognition of its errors forced the loser to make rapid political concessions that allowed for a quick negotiated settlement. Chapter 7 turns to two cases—the 1982 Falklands conflict and the 1919–21 Franco-Turkish War—that were driven by domestic politics. In each case, the domestic political constraints identified in the discussion of the principal-agent mechanism for war prevented the responsible national leaders from escalating the war as they might have wanted to do and ultimately forced a political settlement long before the war produced destruction on the level of the large wars discussed in chapters 3 through 5.

Finally, in the Conclusion, I first recapitulate the central question, main argument, and primary findings of the study. The remainder of the chapter highlights implications of my findings for a number of significant topics, including the study of civil wars, policies for encouraging the political settlement of ongoing conflicts, and the possibility for conflict in the future, both between the United States and China and as a result of the continued spread of nuclear weapons.

[1]

Explanations for Limited and Unlimited Wars

War is, if not common, a persistent feature of international politics. Most wars between countries are, however, limited, lasting days, weeks, or months rather than years, and killing thousands rather than hundreds of thousands or millions. The few conflicts in which intense fighting persists for years, which I call unlimited wars, are thus responsible for a highly disproportionate amount of suffering. I argue that a good explanation for the most destructive conflicts should account both for why these wars did not end more quickly and for why other wars remained more limited. My argument is that unlimited wars are driven by a different mechanism—a different logic of war—from most limited conflicts. In particular, the commitment problem mechanism lacks an internal logic that guarantees that opponents in the war will reach a negotiated settlement after some period of fighting, whereas under the two most prominent alternate logics of war, which I refer to as the informational and principal-agent mechanisms, the revelation of information through fighting ultimately creates a situation in which a mutually agreeable settlement is reached.

In the discussion of the commitment problem mechanism, I distinguish between two types of commitment problems, which I refer to as situational and dispositional. Situational commitment problems are associated with shifting power: in a situational commitment problem, declining powers fear the implications of their decline and thus launch aggressive revisionist wars to prevent the decline from occurring. Given the centrality of preventing decline in this argument, I often refer to this explanation as the preventive war mechanism. The basic logic of this mechanism is quite well understood, although some of its implications are less well appreciated, and even well understood implications have not necessarily been subjected to rigorous empirical tests. Because

[11]

the mechanism is well understood, however, my theoretical discussion is relatively short.

The second—dispositional—type of commitment problem is less well understood, however, and thus merits closer analysis. In this case, leaders decide to fight not because of fear of relative decline but because of a sincere belief that they face an opponent who is by nature committed to aggression, and thus that a viable peace requires at a minimum the replacement of the opposing leadership and possibly a thoroughgoing reform of the opposing society. This belief logically implies that almost any political settlement is unacceptable, with the consequence that only the opponent's unconditional surrender is an acceptable basis for war termination. While the commitment problem logic here is clear, the reason why people come to believe that their opponents are implacably aggressive is less obvious. I argue that these beliefs emerge when the targets of preventive wars misattribute the motivation for the initial attack. In these cases, one side initiates a war out of fear that impending decline will force unpalatable political concessions on it in the future. Their opponents, however, believe that the stated preventive motivation for war is merely rhetorical cover for a war of expansion, motivated not by the situation that confronts the declining power but by the character of the opposing leader, regime, or society. This belief is most likely when the target of the preventive war does not in fact intend to do what the declining power fears. It is this process that produces the refusal to negotiate that is a characteristic of some of the worst wars.

For both the informational and principal-agent mechanisms, the central dynamic keeping war limited concerns information. In informational conflicts, the two sides in the war disagree about the likely consequences of fighting. Events on the battlefield and at the bargaining table confront combatants with developments that contradict their expectations, forcing them to revise their expectations of victory and to accept political concessions that ultimately allow for settlement. In principal-agent conflicts, leaders must lie to their publics to maintain support for the war, but their ability to do so diminishes as others in government and the public more generally observe developments in the war that contradict the leader's claims. The internal limitations in both these mechanisms imply that, while they may account for many of the wars that we observe, they will not be individually responsible for the long, high-intensity wars that produce a disproportionate amount of human suffering.

The chapter begins with a short summary of the bargaining model of war, from which the three mechanisms are drawn. I then turn to explanations for unlimited wars, starting with the situational commitment problem and preventive wars and then turning to the dispositional commitment problem and unconditional surrender. This discussion identifies both general hypotheses about the destructiveness of these conflicts and more specific hypotheses that widen the range of possible empirical tests, ultimately allowing for greater confidence in the general argument. The final section examines the two mechanisms that are associated with more limited wars, focusing in particular on the informational dynamics that ultimately lead to settlement in each case.

THE BARGAINING MODEL OF WAR

The arguments in this book are developed out of what has come to be the dominant framework used in the study of war in the international relations field, the bargaining model of war.[1] This model starts from a few central assumptions about the nature and purpose of war and derives the important implication that most of the time there should be political agreements that both sides in a dispute prefer to the costly gamble of war. From this perspective, a cause of war is something that prevents a settlement from being reached; theory has identified a range of such potential causes.

The core assumptions of the bargaining model are that war is political and that fighting is costly. Both of these assumptions are generally accepted. Thus, for example, Clausewitz started his magisterial study of war with the observation that "war is merely a continuation of policy by other means"; in other words, war is the attempt to gain through force what could not be acquired through diplomacy.[2] But if war is fought to divide a political stake, and if diplomacy could also divide that stake without imposing the costs of fighting, then even people who disagree vehemently about the appropriate division have a strong reason to resolve their dispute without resorting to war. This finding accords with the empirical observation that there are many more political disagreements in the world than there are wars. In contemporary politics, we have disagreements over territory (Israel and Syria or Bolivia and Chile, among many others), the pursuit of nuclear weapons (opposition to North Korea and Iran), economic or environmental policies (e.g., disputes about plan to dam rivers like the Nile or the Jordan), access to natural resources (e.g., disagreements over access to oil and other resources in the South China Sea), the treatment of ethnic minorities (e.g., ongoing tensions related to the treatment of ethnic Serbs, Albanians, Hungarians, and others in Eastern Europe), and many other issues, yet wars arising from these disputes are rare. The puzzle of war, then, is why leaders sometimes fail to agree on a division of the political stake and instead opt to resort to force, with the costs that that decision entails.

[13]

Many logically coherent answers to this question exist. Leaders might choose war because they disagree about the likely result on the battlefield, because they doubt that an agreement today would be upheld in the future, because they simply cannot divide the issue at stake without destroying its value, because they do not suffer the costs of war personally and thus do not share their constituents' preferences, because they are unusually acceptant of risk, because they (or their constituents) actually enjoy the experience of fighting, because they fail to coordinate on an efficient equilibrium settlement, or because the costs of maintaining a suitable deterrent are higher than the costs of war. That said, some of these explanations are empirically more important than others. The remainder of this section thus discusses these different logics, explaining why I choose to focus on the three that are central to my analysis while disregarding the others.

The three mechanisms on which I focus have been the subject of significant scholarly research. The idea that war may happen when leaders disagree about how the war is likely to go draws strength from Blainey's finding that mutual overoptimism has been present in a tremendous number of wars through history; it is thus unsurprising that this explanation for war has received more attention than any other.[3] Similarly, the idea that fears associated with decline may bring about war has a long history, traced back to Thucydides' argument that the Spartans started the Peloponnesian War because they feared the rise of Athens.[4] Arguments about a link between domestic politics and war likewise have a long history, drawing on Kant's argument that war happens because leaders do not suffer its costs, and building on a range of cases in which diversionary motives or other domestic political processes are believed to have accounted for one side's decision to fight.[5] All of the mechanisms that are discussed in this project thus have been the subject of intensive prior inquiry, largely because scholars believe them to be quite important in the real world.

The remaining potential explanations do not meet this standard, and moreover for several there are significant obstacles to their identification in individual cases. Issue indivisibilities likely constitute the most controversial omission from this study. In practice, however, few issues are truly indivisible—even something like control over government can be divided through power-sharing—and even a dispute over truly indivisible issues can be resolved short of war if leaders have recourse to side payments.[6] The stance that indivisible issues are not particularly important has met some criticism, but from an empirical perspective an approach grounded in indivisible issues faces the significant challenge that actors frequently claim that stakes are indivisible but subsequently proceed to divide them.[7] Thus, for example, rebel groups such as the

Free Aceh Movement in Indonesia or the Tamil Tigers in Sri Lanka long insisted on independence as an indivisible and nonnegotiable position only to backtrack to accepting autonomy once war turned against them. Given that people are ingenious enough to find ways of dividing even the most apparently indivisible of goods, in practice convincing demonstrations that indivisible issues prevent settlement are rare and typically arise not because a good is actually indivisible but because it comes to be seen as such.[8] Relatively few examples of this process have been observed, however, and existing examples are drawn almost exclusively from internal conflicts. In a study of interstate wars, therefore, this mechanism can be safely discounted.

A number of other potential mechanisms exist, but all are ultimately problematic. Explanations grounded in a positive utility for the experience of fighting or risk-acceptant preferences both dramatically overpredict the frequency of war and are thus theoretically uncompelling. For example, if war occurred because people enjoyed fighting, we would expect war to be a constant feature of politics, at least wherever those who enjoy fighting are found. From a more social scientific perspective, scholars have typically found explanations for inefficient behavior such as war that are grounded directly in preferences—people fight because they enjoy fighting, or because they like risk—to be theoretically unsatisfying. At base, they explain away outcomes that we believe to be undesirable by claiming that the people involved actually enjoyed them, an unsatisfying and, for an empirical perspective, typically unconvincing approach.[9] In this sort of case, it would be better if the preference were explained by a more general theory, as for example is done later in this chapter by demonstrating that the preventive logic of war provides a rationale to engage in more risk-acceptant behavior. Empirically, I am unaware of any cases in which scholars have made convincing arguments that a preference for war or a simple preference for risk led to war in the modern international system absent other, stronger explanations.

The remaining possibilities also have not been demonstrated to account for many wars empirically. The possibility that difficulties coordinating on an efficient equilibrium might lead to war, while theoretically coherent, relies on the existence of multiple equilibria, making it very difficult to sort out empirically; it is unclear why leaders would ever coordinate on an equilibrium that is nonobvious even for game theorists.[10] Likewise, the argument that the cost of maintaining a deterrent might be high enough to justify going to war is unconvincing at least for the modern era given the extraordinarily high costs of war, especially of a war that would be sufficient to allow a country to abandon the need for significant spending on deterrence.[11]

[15]

EXPLANATIONS FOR UNLIMITED WARS

The central argument in this book is that the most destructive wars are driven by commitment problems. I start by discussing the best-known source of commitment problems, in which a declining power fears that its rising opponent will make intolerable demands in the future and launches a war to prevent the decline from occurring. After discussing the reasons why starting a war under these circumstances may be rational, I examine the ways in which such a war might end, highlighting in particular the absence of reasons why a negotiated settlement should necessarily become more likely over time. This argument, which implies that preventive wars driven by shifting power can be unlimited, has a number of testable implications for the destructiveness of war as well as the conduct of the war and the way in which it is ended.

The second half of this section then turns to a novel commitment problem logic that leads to a categorical refusal to negotiate with one's opponent despite unusually high costs. I argue that this behavior arises when targets of preventive wars conclude that the initiator of the conflict is a war lover who will repeatedly attack absent a fundamental revision of its government or society. This inference occurs when the target of the preventive war does not in fact want to do what the initiator fears; given the inference, no negotiated settlement to which the initiator would ever agree is acceptable, leading to a sincere demand for unconditional surrender. Again, this argument identifies a number of testable implications about when unconditional surrender demands will arise.

Shifting Power and Preventive War

The idea that fear of decline might lead someone to launch a preventive war is certainly not new: Thucydides famously attributed the Peloponnesian War to "the growth of Athenian power, and the fear which this inspired in Sparta," and just about every systemic war in the modern state system has been attributed to one side's belief that time favored its opponents.[12] The rise and decline of great powers lies at the core of the power transition theory of war and of related cyclic theories of international politics.[13] There is thus precedent for believing that power transitions will tend to be associated with great power wars. More recently, work on the bargaining model of war has built on this logic, clarifying the reasons why war might be a rational response to decline.[14]

Historical examples of the preventive motivation for war are easy to find. The American invasion of Iraq in 2003 clearly fits with this preventive logic: convinced that "time is not on our side" and that the danger "only grows worse with time," the government could not afford to wait for definitive evidence of Iraqi duplicity that might come "in the form of a mushroom cloud."[15] Nor is this example unique in American history: one can point to preventive motivations in the American entry in World War II, and several leading military, political, and intellectual figures contemplated preventive war against either the Soviet Union or China after the war.[16] Innumerable examples from other countries and other eras provide further evidence that preventive motives, while often not acted on, are frequently present.

To understand why war might arise out of shifting power, it helps to start from the basic bargaining model observation that even countries with diametrically opposed interests will prefer to resolve their disagreements politically rather than militarily: with the right settlement, both sides will prefer the political status quo to the costly gamble associated with war.[17] If power is shifting, however, the knowledge that your opponent prefers the status quo to war today does not guarantee that she will hold the same preferences in the future. If over time one side is becoming more powerful relative to the other, then as time passes it likely will demand more as a condition for not going to war. Indeed, a significant literature has focused on the role of shifting power over time—largely as a consequence of differential economic growth—as a determinant of change within the international system.[18] Thus, for example, once a unified, powerful Germany supplanted the muddle of small principalities that preceded it, German demands grew to encompass things in which they previously had shown no interest, such as African colonies (Germany's "place in the sun"). This situation need not imply that war would inevitably happen in this scenario; indeed, it may well not. However, once the shift has occurred, the declining side will find that it is in its interest to acquiesce in response to demands that it initially would have rejected. Thus, to continue the example, the other European powers found it prudent not to directly oppose the German acquisition of colonies in Africa and elsewhere, with the result that Germany had assembled a significant empire by World War I. In other words, once the power shift has taken place, the declining power has a choice between unpalatable concessions and war on more difficult terms.

Given the concerns that relative decline generates, the rising power—which prefers to experience its rise unmolested—will often be trying to signal its peaceful intentions as clearly as possible, as for example China appears to be doing today. Yet precisely because the rising power wishes to be left alone until its rise is complete, even hostile rising powers will want to signal benign intentions; this possibility for duplicity means that the declining power may doubt the authenticity of the signal. Moreover,

[17]

in some cases rising powers may be unable or unwilling to signal peaceful intentions, even when an opponent's fears are unjustified. Prior to the Paraguayan War, discussed in chapter 3, the Brazilians did not signal benign intentions toward Paraguay, apparently because they believed their lack of hostility to be self-evident. Prior to World War II, Germany's opponents—in particular the Soviets—failed to successfully reassure Hitler in part because they did not share his view of the world and thus did not recognize the threat that he perceived.[19] And more recently, Saddam Hussein apparently felt unable to signal the absence of a WMD program convincingly prior to the 2003 US invasion without simultaneously signaling weakness to Iran and thus inviting attack.[20]

Thus leaders who anticipate relative decline must decide how to respond, often without knowing for sure what the rising power intends to do once stronger. In this context, there typically are no perfect policy options. If the declining power is confident that the rising power's intentions are benign, or if the anticipated shift is relatively small, it typically will make sense simply to accept whatever limited concessions are associated with decline.[21] Thus, for example, when initial expectations after World War II that Britain would be just as influential as the United States and the Soviet Union proved impossible to sustain, the British came to accept an apparently inexorable decline that saw the liquidation of the largest empire in history and increasing deference to the United States.[22] In some cases, as with the construction of NATO in response to the rising Soviet threat after World War II, it may be possible to build an alliance strong enough to deter core demands even after a shift has occurred. An alliance strategy is risky, however: history is replete with shocking reversals in alliance structure, such as the Molotov-Ribbentrop Pact or the Sino-American rapprochement in the 1970s. Alternately, internal balancing—shifting production from domestic consumption to military power—might allow a declining power to keep pace for a time, but in a competitive environment there may be little room for additional military expansion, while excessive production can have significant economic costs.[23]

From a formal perspective, a deal in which the rising power commits to an indefinite continuation of the status quo—assuming that each power prefers the current status quo to war today—in return for being left alone will be preferable on each side to the expected utility of going to war. The problem, however, is that this agreement is not credible.[24] In effect, the declining power is making an immediate concession—refraining from using force while it is relatively strong—in return for the rising power's promise of future concessions—not taking advantage of its increased strength once its rise is complete. Having received what it wants from the agreement first, however, there is nothing to prevent the rising power from refusing to make its concessions in return; indeed, given the pressures of the international system, it may well be stupid for it not to. An intelligent leader of the declining power, however, must recognize these incentives and thus assume that no bargain of this sort with the rising power will actually work. In the end, then, declining leaders often expect diplomacy to yield the same outcome as simply permitting the rise to occur and accepting the consequent loss of influence.[25]

Given the limitations to diplomatic strategies, in at least some cases leaders will conclude that war is an attractive policy option. Specifically, a big enough victory in war potentially can resolve the declining power's problem by preventing the power shift in the first place. The declining power can aim to acquire militarily strategic territory that would make an attack by even a stronger opponent extremely difficult, or it could acquire territory that is critical for the other side's industry. Thus, for example, one solution that the French pursued in the aftermath of the First World War to overcome the problems associated with the superior German economy and larger German population—which implied that Germany would inevitably recover from its defeat—was to either annex or render independent the Rhineland, which was the territory through which any German attack would have to come and which furthermore was the center of German industry. From a purely military position, the size of the Entente's victory and especially the dissolution of the German army after the signing of the armistice meant that they were in a position to impose such terms; however, opposition from their Allies forced the French to settle for the demilitarization of the Rhineland.[26] Had they been successful, it would have been significantly harder for Hitler to break out of the Versailles straightjacket and turn Germany once again into the sort of country that could threaten all of Europe. In general, if the declining power believes that it still has a reasonable chance of achieving a victory large enough to permit it to impose these kinds of concessions on its opponent, then going to war will be a potentially attractive option. This belief that recourse to arms may forestall the anticipated decline constitutes the preventive motivation for war.

To summarize, declining powers have few good options when faced with a potentially antagonistic rising power. Because of international anarchy, diplomatic solutions are unreliable, while doing nothing invites eventual demands for concessions that the declining power either must accept or must oppose militarily from a position of relative weakness. In this context, forcibly imposing a significant defeat on one's opponent holds out the potential to resolve the entire problem in one quick move. Thus while power shifts need not necessarily cause war, they provide a potent motivation for fighting.

[19]

How Do You End a Preventive War?

In those cases in which the declining power opts for war to prevent decline, however, they create a conflict that is unusually difficult to bring to an end. For war to end, it is necessary that the problem that led to fighting be resolved. Thus, in this case, for the two parties to agree to a settlement it must be the case that the motivation to fight a preventive war either disappears or diminishes to the extent that disincentives to fight trump the desire to prevent the decline from occurring. As I discuss below, there are ways in which the motivation for preventive war might diminish or disappear, but none are guaranteed to occur. Indeed, in most formal models of war, when shifting power brings about war, the participants end up fighting to a final military outcome in lieu of reaching a political settlement.[27]

For the war to end, it must be the case that the power that anticipates decline either be unable to continue to fight or see fighting as no longer in its interest. In a basic formal model of bargaining within a commitment problem war, a round of fighting either renders one side in the conflict militarily victorious over the other or leaves the actors confronting the exact same situation as before, having simply suffered the costs of fighting. In this sort of model, if fighting was rational in the initial stage, it will remain rational in any subsequent stage (as the costs of fighting are sunk and should not influence prospective decisions), implying that commitment problem wars will only ever end when one side loses the ability to continue to fight.

In reality, of course, battles can change the relative situation of the participants without inflicting a decisive defeat on either side. The question is how much change would be necessary for settlement to be reasonable. Typically, addressing the sources of relative decline will require a major revision to the status quo, such as alienating productive, populous, or militarily strategic territory from an opponent's control, replacing an existing government, or greatly weakening an enemy's military. Thus, for example, the German fear of increasing Russian power prior to 1914 could be assuaged only through a dramatic revision of territorial control in Eastern Europe, as ultimately happened in the Treaty of Brest-Litovsk, in which the Russians ceded Finland, the Baltic States, Poland, Belarus, and Ukraine to German control or annexation. In the Crimean War, discussed in chapter 5, the British believed that preventing the Russians from gaining control over the Black Sea Straits, which would have uncorked the Russian fleet in the Mediterranean and thus threatened British overseas interests, required that Russia cede most of its territorial gains around the Black Sea from the previous hundred years.

[20]

Such expansive war aims will be hard or impossible to achieve through a single battle. In 1914, the first major battle on the Eastern Front—the surprise German victory in the Battle of Tannenberg—did little to address the sources of Russian rise, which hinged on the potential power implicit in Russia's population as well as military recovery from the significant defeat in the Russo-Japanese War. It was only after several years of persistent defeats that the Russians could be convinced to make the concessions that the Germans saw as necessary for peace.[28] Absent such a victory, an end to the war on intermediate terms would simply have returned Germany to the strategic situation that had existed prior to fighting, when war was seen as preferable to peace.

This point should not imply that settlement is impossible. Large victories may address the concerns about decline that initially motivated the war, as the Eastern Front in World War I demonstrates. Alternately, a series of victories may put the rising power in a position to impose today the concessions that the declining power was fighting to prevent, or may render it impossible to achieve the victories necessary to prevent the rise from occurring. Once such a point is reached, there is nothing to be gained from further fighting, and the defeated declining power logically should be open to settlement. In the Paraguayan War, for example, Francisco Solano López demonstrated increased willingness to negotiate once repeated defeats demonstrated that Paraguay was not going to achieve the significant victories he had initially sought. A similar situation could logically arise if the feared shift occurs in the course of the war, as, for example, in a case in which the target of a preventive war demonstrates that it has acquired the nuclear arsenal that the war was designed to prevent it from getting. Another logical possibility is that if fighting destroys the value of the good over which participants are fighting, settlement may eventually become possible as the future concessions become relatively painless.[29]

Any of these scenarios could potentially arise, and indeed in some cases may arise relatively quickly. It is also, of course, possible that one side in the conflict quickly achieves a decisive military victory, as with the American conquest of Iraq in 2003, which allowed the Americans as occupiers to address their concerns about the Iraqi pursuit of nuclear weapons.[30] None, however, are *necessary*: if the military course of the war is such that the declining power fails to achieve the revisions necessary to prevent the decline from occurring and yet still believes those revisions to be possible, it is likely that they will wish to continue the war. Moreover, the expansive war aims that are typically required to prevent the feared decline imply that even fairly significant military victories may be insufficient to address the commitment problem concerns. At some point, of course, the participants will exhaust themselves—in the limit, everyone will be dead—but well before this point is reached the war will have graduated from the limited to the large.

It is important that I be explicit about one aspect of this argument. I am not claiming that preventive wars are necessarily unusually destructive—cases like the conventional phase of the 2003 Iraq War provide obvious counterexamples to such a hypothesis. Instead, my argument is that there is nothing in the logic of this mechanism that *prevents* these wars from being so destructive. Figure 1.1 illustrates this argument graphically: in an interstate context, commitment problem wars will be fought intensely, but may not be long if one side is able to achieve a quick and decisive victory.[31] By itself, therefore, this argument implies only that the commitment problem mechanism provides a logically coherent explanation for unusually destructive wars. The stronger argument that unusually destructive wars will tend to be driven by the commitment problem mechanism thus is incomplete without the argument, which I advance in the final section of this chapter, that alternate logics of war are necessarily limited. In this sense, the logic behind the core argument that commitment problem wars will tend to be unusually destructive is as yet incomplete; I nonetheless present that argument here while acknowledging the need for further discussion of other mechanisms.

Hypothesis 1: Preventive wars will frequently be unusually destructive.

In addition to this central prediction, a closer examination of the logic of preventive wars uncovers a number of additional hypotheses; while the most important predictions in this book concern the destructiveness of

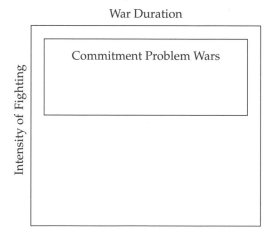
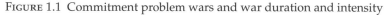

FIGURE 1.1 Commitment problem wars and war duration and intensity

war, support for these ancillary hypotheses should increase our confidence in the explanation for particularly destructive wars by demonstrating that the preventive war mechanism is operating in the manner that theory predicts. Thus one relatively straightforward implication is that in preventive wars the declining power will frequently have unusually large war aims.[32] Addressing the preventive motivation, as noted above, will typically require substantial modifications to the status quo that address the source of the feared rise; absent such success, the problems that led to war in the first place will not be resolved. This requirement is a reason why such wars will tend to be rare, but it also provides a rationale for relatively large war aims, as with the German desire in 1914 to make major territorial gains at Russia's expense. The contrast between American war aims in the Persian Gulf War and the 2003 invasion of Iraq is also informative. In the former case, as discussed in chapter 6, the informational mechanism accounted for fighting, and the United States, although committed to limiting the threat Iraq posed to its region, restricted its war aims by refraining from deep drives into Iraq or direct attempts to topple Saddam. In 2003, however, the fears associated with the belief that Iraq was nearing acquisition of nuclear weapons could be addressed only through occupation and regime change, hence the coalition's willingness to pursue goals that had been seen as too costly or risky twelve years previously.

> Hypothesis 1a: In conflicts driven by the preventive war mechanism, the declining power's war aims will tend to be unusually large.

These large aims, however, are often beyond what the declining power would expect to achieve in a typical war. Thus, for example, in the Paraguayan War case discussed below, a fear of future decline led Francisco Solano López to launch a highly aggressive war against both Brazil and Argentina, the great powers of South America; although he had some reasons for optimism, such a war would not typically be seen as a good gamble for a buffer state like Paraguay. Leaders thus will frequently need to resort to unusually risky war plans to have a reasonable chance of achieving such goals. In the Paraguayan War, these risks manifested as deep incursions into hostile territory, where elite forces were at risk of being cut off and forced to capitulate, as well as a risky diplomatic strategy of provoking war with Argentina in the hope that several disaffected Argentine provinces would rally to Paraguay's side. Similarly, the conventional interpretation of German strategy in World War I highlights the riskiness of the Schlieffen Plan, in which the Germans left the eastern frontier weakly defended while striving to knock France out of the war in the first six weeks; the subsequent shift in 1917 to unrestricted

[23]

submarine warfare was also recognized to involve great risks for Germany.[33]

> Hypothesis 1b: In conflicts driven by the preventive war mechanism, the declining power will tend to adopt unusually risky strategies.

All the previous hypotheses in turn imply that conflicts driven by the preventive war mechanism will tend to end in military conquest rather than a negotiated settlement. Most obviously, resistance to voluntary war termination—negotiated settlement—makes involuntary war termination—conquest—more likely. The recourse to risky strategies amplifies this point, as the risky strategies that the declining power is willing to adopt may produce decisive military breakthroughs or catastrophic defeats. Relatedly, these arguments imply that the increase in the destructiveness of preventive wars will be primarily a consequence of the difficulty participants face identifying a mutually agreeable settlement rather than anything about the speed with which one side conquers the other militarily; indeed, if anything, these wars should manifest faster conquest.

> Hypothesis 1c: Conflicts driven by the preventive war mechanism will be unusually likely to end through military conquest as opposed to settlement.
> Hypothesis 1d: Preventive wars will take an unusually long time to reach settlement, but will if anything be associated with quicker military conquest.

Before moving on, I need to briefly distinguish among several different ways in which shifting power might be associated with war, only some of which are likely to produce these sorts of high-cost, high-risk conflicts. The most obvious scenario is one in which one country is consistently rising in power relative to another, as with the rise of China relative to the United States today. Alternately, one country's sudden decrease in relative capabilities may create a temporary window of opportunity for an opponent, as with purges after the Iranian Revolution that weakened Iran relative to Iraq. A third possibility covers situations in which control over a particular good—most frequently strategically located territory—could generate a discontinuous shift in relative capabilities. The Crimean War provides a pertinent example: the British feared that Russian acquisition of Constantinople would dramatically shift the regional distribution of power by allowing Russia to exclude enemies from the Black Sea and to project power into the Mediterranean. All of these situations produce incentives to engage in the sorts of preventive war described here.

By contrast, two other scenarios in which anticipated power shifts may influence policy do not produce the dynamics that I describe here. Preemptive wars, in which one side begins a war to take advantage of first-strike advantages because it expects its opponent to attack in a matter of days, logically involve shifting power—the first-strike advantage—in the decision to start the war. That said, preemptive wars are empirically rare and occur only when war was already very likely to occur.[34] The way in which the war is fought thus depends on what mechanism brought the opponents to the brink of war initially: preemption in the context of the preventive motivation for war may produce a large conflict, but preemption when the initial motivation for fighting is disagreement about relative strength or relative resolve is likely to be associated with a more limited war. Second, sometimes developments within a war create an anticipation of future military shifts, for example because one side's army is vulnerable. In this situation, the side with a military advantage will have an additional incentive not to settle until after the anticipated defeat has been imposed. Thus, for example, once the Soviets encircled the German Sixth Army at Stalingrad in World War II, they would have resisted a settlement that allowed the Germans to escape. Similarly, the seizure during war of strategic territory whose full utilization would affect the future balance of power, as with the American seizure of the Philippines in the Spanish-American War or the Iraqi control of Kuwait (and hence Kuwaiti oil) in the Persian Gulf War, can complicate negotiations by rendering some possible agreements unsatisfactory, but leaders will be much more willing to negotiate away these advantages when they are not the central problem at stake in the war. While these sorts of developments may marginally lengthen a war, the broader dynamics of the war will again depend on the mechanism that led to fighting in the first place.

Commitment Problems, Evil Dictators, and Unconditional Surrender

The previous section argued that a leader who believes that a hostile opponent is rising in power relative to her country at times will decide to launch an aggressive war that can end up being extremely destructive. This preventive war mechanism constitutes what I refer to as a situational commitment problem: because the opponent is rising in power, the declining power cannot trust its promises not to seek political concessions at a later time. In some cases, however, the concern that an opponent cannot commit to the status quo is grounded in a different logic: rather than mistrust arising out of a logical response to shifting power, leaders simply conclude that the opponent is by nature (i.e., dispositionally) committed to war. As I discuss below, this sort of belief is at odds with the conventional bargaining model of war, and existing studies have not deeply analyzed either the sources of such beliefs or their implications for foreign policy.

In this section, I examine dispositional commitment problems in greater detail. I first discuss why the belief that one's opponent is dispositionally committed to war is problematic for standard bargaining model approaches. I then turn to the implications of such beliefs for wars in which they arise. If one side sincerely believes that war is a consequence of an opponent's innately aggressive character, such that any peace agreement will merely set the stage for a new attack, then the only path to a viable peace is the reformation or removal of the offending actors on the opposing side, be they an individual leader, a broader government, or even the entire society of the opposing country. In practice, this sort of goal can only be achieved through a total military victory, rendering any possible political settlement unacceptable. Thus when leaders conclude that their opponent is dispositionally committed to war, they categorically reject negotiation, even in the face of high costs and uncertain military outcomes. I refer to this behavior as a sincere demand for unconditional surrender.

The remaining question, however, is where this sort of belief comes from. While one could imagine a variety of different possible sources, I argue that interstate wars in which this sort of behavior has arisen have followed a common mechanism. Specifically, a declining power launches an aggressive preventive war, based on the belief that its rising opponent will impose painful concessions on it once its rise is complete. In some cases, however, the rising power lacks the intentions that the declining power ascribes to it and as a result often fails to appreciate the true motivation behind the declining power's aggressive war. In short, confronted by an opponent who justifies an aggressive war by reference to a threat that the target of that war knows to be nonexistent, the target concludes that the stated justification for war is merely cover for some other motivation. Given this disjuncture, leaders may reasonably conclude that only a character-based explanation can account for their opponent's aggression. In summary, a declining power launches a preventive war in response to a feature of its environment—relative decline with respect to a presumed enemy—but that enemy misinterprets the attack as evidence of a dispositional commitment to aggression that in turn poses an insuperable obstacle to any possible political settlement. I thus argue that this process, which links situational and dispositional commitment problems, is responsible for the most destructive interstate wars.

[26]

Stage Two: From Evil Dictators to Unconditional Surrender

Political scientists have noted the existence of the dispositional commitment problem. Thus, for example, Reiter attributes the British refusal to negotiate with Germany in the summer of 1940 to the fact that they simply "did not trust Hitler to adhere to any war-ending commitment," not because Germany was growing stronger but because Hitler was fundamentally untrustworthy.[35] To date, however, no one has provided a convincing explanation for the belief that an opponent cannot be trusted to live up to *any* possible peace agreement.

Indeed, such a belief is at odds with the basic logic of the bargaining model to a degree that has not been fully acknowledged. The best extant explanation for such mistrust is that it is a consequence of the uncertainty associated with international anarchy.[36] On closer examination, however, this argument is unsatisfying. Setting aside situations of shifting power (situational commitment problems), and assuming that other mechanisms for war are not in play, mistrust alone should never provoke war. Mistrust is the assumption that one's opponent's intentions are hostile when they may not in fact be so. Even if we assume the worst-case scenario of maximally hostile opponents (i.e., those with perfectly opposed preferences over the political stakes), the bargaining model predicts that given the costs of war each should prefer a political settlement along the lines that would be reached through fighting rather than going to war. In other words, the worst assumptions about your opponent's interests should not mean that you are unwilling to negotiate. In these cases, however, as I discuss further below, the mistrust is associated with a categorical refusal to negotiate.

That said, certain types of actors might theoretically fit the depiction in the dispositional commitment problem. An opponent who was undeterred by the costs of war, for example because she enjoyed the experience of fighting, might be expected continually to start new wars, even in the absence of any standard war-producing mechanism. Alternately, if one side in a war is by personality congenitally overoptimistic, such that it consistently holds reasonable political proposals by its opponents to be worse than what could be achieved through war, then that power might consistently launch new wars.[37] International relations theorists generally resist including such actors in their theories, as such actors would never cease fighting—a prediction that is at odds with the observation that even the most aggressive states have been open to political compromise.[38]

Yet in rare historical cases people do come to believe that they are faced with such an opponent, even if (as I argue below) the opponent's motivations are frequently more complicated. Thus, speaking of the main Axis powers in World War II, Franklin Roosevelt averred unless disarmed "they will again, and inevitably, embark upon an ambitious career of world conquest," at least unless the Allies forced them to "abandon the philosophy which has brought so much suffering to the world."[39] Similarly, in the Paraguayan War the Brazilians justified their refusal to negotiate with the claim, apparently sincerely believed, that they were facing a tyrant with unlimited ambitions; in the Iran-Iraq War the central problem from the perspective of the Iranian leadership was that their opponents were "corrupt."[40]

The implications of this sort of conclusion are stark, as is apparent if we return to Blainey's dictum that any cause of war must have an associated cause of peace, or in other words that for a war to end, the problems that led it to begin must be resolved.[41] If your opponent's evil nature is the cause of your war, then peace cannot be restored so long as that evil nature is a concern.[42] In general, this sort of observation has tended to lead IR scholars to discount rhetoric about the evils of the opponent: leaders always allude to evils on the other side, but most wars end without fundamental political change in either side's regime. But this point should not blind us to the implications that arise when one side concludes that war is a consequence of the opponent's character. If evil dispositions are a cause of war, then fundamental change will be needed for peace.

What is required to solve this problem will depend on how deep-seated the commitment to aggression is believed to be. It is conceivable, especially in more personalistic dictatorships, that the problem would be believed to lie solely with the individual leader, in which case the replacement of that leader might be sufficient to bring peace. In the Paraguayan War, discussed in chapter 3, it is quite possible that the Brazilians saw Paraguay in these terms. In other cases, however, all members of the governing regime, the military, or even the entire population of the country may be implicated, as was the case for Allied leaders trying to determine what to do about Germany. In this case, viable peace will require more thoroughgoing reform. In any case, however, leadership change is going to be seen as a nonnegotiable prerequisite to peace. That requirement in turn makes a mockery of negotiation with the existing government, and thus justifies a stance of simply refusing to negotiate so long as that government is in power. To the extent that the source of the aggressive disposition is believed to lie deeper in society, negotiation with any possible interlocutor will be similarly unacceptable, and thus the only viable peace will be one that permits complete reform of the offending country. Thus from a theoretical perspective there is a logical connection between the belief in the dispositional commitment problem and a refusal to negotiate: if you are unwilling to accept any peace deal that could possibly be accepted by the opposing government, what purpose is there in holding talks?

In short, if leaders on one side believe that war follows from the character of the opposing leader, government, or society, then the only reasonable policy response is a war explicitly fought for regime (or even societal) change, in which they refuse to countenance any form of negotiation. I refer to this behavior as a sincere demand for unconditional surrender. By "unconditional surrender," I refer to a demand that the opponent's military surrender in its entirety and that the opponent accept complete loss of control over territory and government. As Franklin Roosevelt said during World War II, unconditional surrender meant that he was "not willing at this time to say that we do not intend to destroy the German nation."[43] This stance thus effectively forecloses political settlement as an option for ending the war.

The focus on *sincere* demands for unconditional surrender highlights an additional requirement: that the side demanding unconditional surrender be willing to stand by this demand even in the face of high costs and an uncertain outcome from continued fighting. References to the iniquity of one's opponent are common in war, as are claims that one will never negotiate. In most cases, however, leaders are not willing to stand by such demands when costs are high or when victory is uncertain. In some cases, as for example in the latter phases of the Russo-Hungarian conflict of 1956 or at points during the American invasion of Iraq in 2003, leaders refuse to negotiate because they believe their side to be so overwhelmingly militarily dominant that limited fighting will permit the dictation of terms without negotiation.[44] In these sorts of cases, available evidence suggests that unexpected military setbacks will be associated with a new openness to negotiation, something that one does not see in cases of sincere demands for unconditional surrender. This definition also excludes limited military demands for unconditional surrender, as with American demands that Spanish forces on Cuba (but not elsewhere) surrender unconditionally during the Spanish-American War, or the British demand that Argentines occupying the Falkland Islands (but not those on the mainland) surrender unconditionally during the Falklands War. Limited unconditional surrender demands arise quite frequently within the context of war and do not follow the logic discussed here. Instead, a sincere demand for unconditional surrender constitutes a refusal to contemplate a negotiated end to a war, even in the face of high costs in the pursuit of uncertain military victory.[45]

Hypothesis 2: Dispositional commitment problems—the belief that one's opponent is by nature committed to aggression—will be associated with

[29]

sincere demands for unconditional surrender, which produce wars to the death.

Sincere demands for unconditional surrender are thankfully rare—most wars end in a negotiated settlement, typically fairly quickly, and even in extended wars leaders are typically open to some political settlement, if not one that their opponent would also accept. Indeed, by the standards described above, only four interstate conflicts since 1816 fit this description: the nineteenth-century Paraguayan War, the Pacific War in World War II, the European War in World War II, and the Iran-Iraq War.[46] At a per capita level, however, this list of wars comprises some of the bloodiest conflicts over the past two centuries, as might be expected from what are effectively wars to the death.

Stage One: From Preventive War to Evil Dictators

The discussion until now has assumed the existence of a sincere belief that one's opponent is irrevocably committed to aggression. As was noted previously, however, the existence of such an opponent seems at odds with the assumptions of the bargaining model. More important, these beliefs simply are inaccurate. Franklin Roosevelt may have sincerely believed that Hitler was planning to invade the Americas via Brazil, but subsequent historical studies—many intended precisely to demonstrate Hitler's global intentions—have turned up no credible evidence of such intentions.[47] Similarly, whatever the Brazilians believed about Francisco Solano López's worldview, he was consistently willing to discuss peace with outside mediators, who complained more about Brazilian intransigence than about any Paraguayan commitment to war.[48] The question, then, is where the belief in the dispositional commitment problem comes from.

While there are a variety of ways in which this belief might arise, I argue that a similar process drove all of the cases discussed here. Specifically, demands for unconditional surrender have arisen historically when the targets of preventive war have misinterpreted the motivations behind the attack. Attacked by an opponent who fears future decline, leaders attribute the attack to an innately aggressive disposition. This section details the reasons why one might expect such a misattribution to occur, as a prelude to any attempt to draw specific hypotheses about the circumstances in which it is more or less likely.

The first step in this process is preventive war. Indeed, preventive motivations played a significant role in the initiation of all wars in which demands for unconditional surrender were made. This connection makes sense: as I argued above, in wars driven by a preventive motivation, the initiator tends to have particularly high war aims and is willing to adhere to those aims even in the face of initial military defeats. By contrast, informational and principal-agent conflicts are internally constrained: initial war aims are typically more limited, and leaders either choose to or are forced to scale back their demands relatively quickly, especially if the war goes poorly.[49] From the perspective of the target of a preventive war, however, large or poorly expressed war aims that do not seem closely bound to military developments are particularly compatible with a view of the opponent as a war lover. Not every preventive war involves this sort of misunderstanding, however. Thus the British preventive motivation in the Crimean War—to prevent Russia from acquiring control over the strategically significant Black Sea Straits—was basically understood in Moscow, with the result that the Russians remained open to negotiation. It is thus worth exploring in greater detail the reasons why such a misinterpretation might occur.

> Hypothesis 2a: Sincere demands for unconditional surrender occur in wars in which preventive motivations provided the primary reason to fight, but the demand will be made by the *target* of the preventive attack.

A rationalist explanation would start from the assumption that leaders believe that a small but nonzero proportion of other leaders are war lovers who will continue to launch aggressive wars until they are removed from power. Given that the behavior of some past leaders—most obviously Hitler—is popularly interpreted in precisely this way, such a belief would be reasonable, even if it accords poorly with the basic logic of the bargaining model. Starting from these prior beliefs, leaders would then revise their understandings of their opponent in response to the opposing leader's policies. Given the low prior probability that any particular leader is a war lover, simply being attacked would not be sufficient evidence to conclude that one's opponent was a war lover. By contrast, attack by an opponent who espouses grandiose war aims—especially relative to her military capabilities—and who contends that those war aims are justified by the prospect that you will engage in actions that you do not in fact intend to undertake will be seen as stronger evidence of a dispositional commitment to aggression.

The key element here is the disjuncture between one side's fears and the other's intentions: an attack justified by fear that you will do something you do not intend to do will seem far more unreasonable than one justified by fears that are in fact correct. Thus the central prediction is that sincere demands for unconditional surrender—which follow from

[31]

belief in a dispositional commitment problem—are more likely when the target of a preventive war does not intend to do what the initiator of the war fears it will do.[50]

> Hypothesis 2b: Sincere demands for unconditional surrender are more likely when the target of a preventive war does not harbor the intentions ascribed to it by the initiator.

This argument, however, just redirects the question to how such an unnecessary preventive war might arise: under what circumstances might we observe a leader launching an aggressive war to prevent something that in fact was not going to happen? The security dilemma—the problem that actions taken to advance one's own security may inadvertently threaten others—provides the most likely answer to this question.[51] The critical point here is that given uncertainty about intentions, each side may undertake actions that are basically defensive in nature but that appear to its opponent as evidence of aggressive intent. Indeed, people frequently fail to appreciate the degree to which their intentions are unclear to others. For example, in the early Cold War, the Soviets installed undemocratic communist governments in Eastern Europe, a move that they believed provided the only reliable guarantee of a friendly buffer against a future invasion from the West; Western powers, however, saw the move as evidence of Soviet expansionism. Both sides saw the other's stance as threatening, in large part because they failed to understand the other side's reasoning.[52] The security dilemma is particularly salient in the context of shifting power, as rising powers have strong incentives to conceal any hostile intentions while they are still relatively weak, making it difficult or impossible for a rising power with genuinely benign intentions to make those intentions clear. In this context, a declining power may mistakenly come to believe that a rising neighbor with benign intentions is in fact hostile and may come to believe that war is an appropriate response to such hostility. The target of such an attack, knowing that its intentions are benign, could be forgiven for putting little credence in the declining power's stated fears.[53] As I discuss in chapter 3, this dynamic was likely at play in the Paraguayan War, in which Brazilian leaders appear to have been completely unaware that Paraguayans might view what appeared to be quite expansionist policies as threatening.

If this argument is correct, an additional implication is that interpretations of the opposing side's motivations will change in response to its preventive policies. In other words, rather than always believing the opponent to be dispositionally committed to war, leaders will come to this conclusion in response to the opponent's aggressive expansion. Thus, for

[32]

example, while Churchill consistently warned about the danger posed by Hitler's Germany, the logic behind that warning changed over time, from a basically Realist argument that a stronger Germany would naturally expect some political concessions to a dispositional argument that Hitler was uniquely aggressive.

> Hypothesis 2c: In cases of sincere demands for unconditional surrender, the conclusion that the opposing leadership or regime is ineluctably aggressive will develop in response to the opponent's preventive policies.

To summarize, I argue that leaders demand unconditional surrender when they conclude that they have been attacked by a country whose leadership is dispositionally aggressive and hence would attack again if given the chance. The context for this development is preventive war, in which the initiator attacks out of the belief that war today is preferable to permitting a presumed-hostile rising neighbor to complete its rise and then demand concessions or fight a war from a position of relative strength. The target of that war may in turn attribute this attack not to the situation of shifting power but to a predisposition toward aggressiveness, thus producing the belief that justifies unconditional surrender. This inference is in turn more likely, I argue, when the target of the preventive attack does not actually harbor the hostile motivations ascribed to her by the rising power. Given the rarity of sincere demands for unconditional surrender, the hypotheses that follow from this argument cannot be tested quantitatively.[54] It is possible, however, to test them through careful case studies, as I do in later chapters.

EXPLANATIONS FOR LIMITED WARS

I thus argue that commitment problems, under two different guises, produce particularly destructive wars. This argument gains strength, however, to the extent that other mechanisms that can bring about fighting *cannot* account for wars that are both long and intense: given that commitment problem wars can be either short or long, depending on whether one side is able to achieve a quick and decisive military victory, the argument that the commitment problem mechanism provides the primary explanation for the most severe wars is credible only to the extent that alternative logics can account only for limited wars. This section thus examines the two primary alternative mechanisms: overoptimism arising out of private information, and principal-agent problems in domestic politics. Both provide a credible explanation for why war would begin, but I argue that in both the revelation of information from in-trawar diplomacy and events on the battlefield ultimately will lead to a settlement before the war becomes unusually destructive.

Limited Wars I: Overoptimism

Overoptimism on at least one side about the probability and ease of victory is easily the most common nontrivial feature of the start of war.[55] It is easy to understand why overoptimistic leaders might demand too much at the bargaining table and hence conclude that the use of force is preferable to the bargain that the other side is willing to accept. As Geof-frey Blainey notes, if disagreement over the likely outcome of war leads to fighting, then in general a necessary condition for war termination will be that the two sides come to agree. Indeed, the bulk of work on ra-tionalist explanations for war has focused on this hypothesis, which I refer to as the informational mechanism.[56]

One potential explanation for long wars, then, is that for some reason the participants took a particularly long time to change their beliefs. In practice, however, I argue that overoptimism arising from private infor-mation cannot reliably explain long wars. A number of studies have noted that fighting reveals information, forcing leaders to update their expectations and revise their demands; these revisions lead to conver-gence in political demands and thus provide the basis for peace. Indeed, I argue that this process will generally happen quite quickly, especially when fighting is intense. This argument implies that informational wars must be limited in either duration or intensity.

Divergent Expectations and the Decision for War

The basic intuition of the informational mechanism is straightforward: if leaders on both sides believe that victory will be achieved easily, then they will prefer war to what their opponent will be willing to con-cede at the bargaining table.[57] Because each side's bargaining position is a function of its expectations, we would expect that when both sides think that they will win, their demands will differ substantially, and thus may be mutually incompatible even considering that a prewar deal has the benefit of avoiding the costs of war.[58] Indeed, it is not even neces-sary that both sides think that they can win, so long as their predictions for the course of the war differ markedly. Thus, for example, in the Viet-nam War, the North Vietnamese did not necessarily believe that they could inflict a decisive military defeat on the Americans, but they did believe that American demands were based on an overestimation of the ease with which they would be able to defeat the Viet Cong militarily.

[34]

For this reason, they were unwilling to make the concessions that the Americans would have demanded as a condition for the end of the war; ultimately, the Americans abandoned their demands and withdrew even though their military was undefeated and remained by most standards far superior.[59]

Divergent expectations can arise from a wide range of possible sources. At the simplest level, participants may simply disagree about the likely course of events on the battlefield. The conventional view of World War I, for example, holds that both sides believed that their superior militaries would permit them to punch through the enemy defenses and impose a decisive defeat in short order. Similarly, in the 2003 Iraq War, confidence in the ease with which Saddam Hussein could be overthrown (and the anticipated positive implications of regime change in Iraq) led American policymakers to believe that even quite revisionist demands were appropriate, while Saddam's confidence that he could embroil the invaders in a costly urban war was one basis for his decision to stand firm.[60] Military strategy provides another potential basis for divergent expectations about battlefield prowess. German generals drew confidence prior to their stunning defeat of France in 1940 in part from the knowledge that their opponents were expecting an attack at the wrong point and thus that there was a chance for an immediate and crushing breakthrough.[61] In 1967, the Israelis were confident that a first strike—which their opponents did not expect—could destroy the Egyptian and Syrian air forces, dramatically shifting the balance of capabilities in that war.[62] And in 1991, the first Bush administration gained confidence, and increased its demands, as it became apparent that a flanking attack west of Kuwait would allow coalition forces to avoid the extensive frontal defenses that the Iraqis had established. In all of these cases, knowledge of one's strategy provided a reason for optimism that could not be credibly conveyed to the opponent without prompting the opponent to take countermeasures that would eliminate the advantage.

Relatedly, in some cases the two sides disagree about the likely behavior of external actors. Thus, for example, the German willingness to attack France and Russia in 1914 is often credited in part to miscalculations about British intentions.[63] Similarly, in both 1848 and 1864 Denmark adopted an aggressive bargaining position in a dispute with Prussia, in large part based on an overoptimistic reading of the probability that the British would intervene on their behalf.[64] More recently, as detailed in the case study of the Persian Gulf War in chapter 6, one reason why Saddam Hussein believed that an invasion of Kuwait in 1990 was a worthwhile gamble was the ultimately erroneous expectation that the Saudis would not permit American forces to operate out of their territory. As with disagreements about relative capabilities, divergent expectations about the likelihood of external support can lead to drastically different beliefs about the likely outcome of fighting, which in turn provides reason for each side to believe the other's political demands to be unjustified.

Finally, expectations may diverge because of misperceptions about relative resolve, with one side underestimating the importance of the issue at stake for its opponent and hence the opponent's willingness to fight, either at all or for an extended period of time. Formally, an actor's resolve will be influenced both by her valuation for the stake—how much she cares about the issue in dispute—and by her valuation for the likely costs of war: the more she values the stake, and the less importance she places on the suffering imposed by war, the more resolved she is. Leaders could miscalculate about either of these components to resolve. Thus, for example, in the Anglo-Iranian War of 1856–57, which I examine in chapter 6, the Iranians recognized that they could not win a war against Britain but believed that changes in the strategic environment meant that the British no longer had any reason to care about the independence of Herat: the British were believed to be irresolute because their valuation for the stake was believed to be low. Alternately, there are cases in which one side miscalculates the willingness of the other to absorb punishment. Thus, for example, one reading of the Vietnam War is that the United States dramatically underestimated the willingness of the Viet Cong to absorb incredible amounts of punishment without capitulating on the political issue at stake.[65] More recently, Saddam Hussein appears to have relied in both 1991 and 2003 on the expectation that Americans would not have the stomach for the deaths on both sides that would accompany urban warfare, an estimation that likely followed from the perception of American, and more generally democratic, casualty aversion.[66]

The bases for disagreement about the likely outcome of a war typically lie in the different information available to the two sides: each knows more about its own capabilities, resolve, and strategy than it does about its opponent's. For this reason, I refer to this explanation for war as the informational mechanism.[67] As expectations diverge, leaders on both sides may come to believe that war is preferable to any negotiated resolution to their dispute that the other would be willing to accept. Once this conclusion is reached, war becomes a logical choice.

Overoptimism and Reality Once War Begins

If divergent expectations account for the decision to go to war, then what happens once the war begins? Prior to fighting, negotiations are complicated by the incentive an irresolute or weak actor has to claim to be resolute and strong to get a better settlement. Leaders may lie. War, however, does not. Once fighting begins, the two sides' divergent expectations are put to the test. A leader who adopted an aggressive bargaining position out of the belief that her opponent lacked the resolve to fight can be presented with immediate disconfirmatory evidence when the war begins. If the two sides disagree about whether or not a third power will intervene in the conflict, at least one is likely to be surprised by that power's behavior when the request for assistance goes out. And if the two sides disagree about the relative quality of their fighting forces or of the strategies that those armies implement, events on the battlefield necessarily must be at odds with at least one side's expectations. In Blainey's words, when both sides are optimistic about the likely course of conflict, "war itself . . . provides the stinging ice of reality" that forces expectations to converge until the two sides' beliefs are close enough to permit a settlement.[68]

In building the intuition for this argument, it is useful to contrast it to an earlier claim that new information such as unexpected battle results might lead each side to adjust its expectations equally, so that the loser might make concessions that previously would have ensured peace but still end up fighting when the winner no longer is willing to accept those terms.[69] Where these two arguments diverge, and where the earlier argument is flawed, is in the source of divergent expectations, which likely concern exactly the things that are revealed in war. The loser of a battle is likely to be more surprised by its results than the winner, and thus will update her beliefs to a greater extent, meaning that expectations, and hence political demands, are closer after the battle than they were before. Even in those cases in which success in battle surpasses the victor's prior expectations, she will still end up updating her expectations by less than her opponent, as the opponent is substantially more surprised. Thus, in some cases, as with the Israelis in 1967, one side's demands may be greater at the end of the war than they were at the beginning, but overall convergence still occurs as the defeated opponent scrambles back from demands that battlefield events have proven to be entirely unachievable.

Updating is neither instantaneous nor perfect, of course. The knowledge that NATO was willing to carry out a bombing campaign against Serbia in the Kosovo War did not necessarily give Slobodan Milosevic insight into whether his opponents would be willing to launch a ground war. And new private information is undoubtedly generated within war, as with the British development of the tank in World War I or the success of the Manhattan Project to construct the first nuclear weapons in World War II. On average, however, the trend is toward settlement. If a

[37]

bombing campaign is not proof of complete resolve, it certainly belies any prior belief that the opponent is completely unresolved. Likewise, there is no reason why new private information will consistently be good. For every technological innovation in war, there are programs like the Nazi search for a superweapon in World War II that deliver less than was expected. In other cases, governments have learned before their enemies that their publics are more war weary or their armies less loyal than previously believed. For new private information to prevent convergence of beliefs, it would have to be consistently positive to an extraordinarily improbable degree so as to offset the inevitable disappointments on the battlefield and at the negotiating table.

The Franco-Austrian War of 1859, in which France aided Sardinia-Piedmont in seizing Austrian territory in Italy, provides a useful example of this process in action.[70] Both sides in this war expected to win quite quickly. The Austrians initially expected to overrun Piedmont before the French could supply effective assistance, and they thus expected to humble the upstart Sardinians and secure their position in Italy; their opponents expected victories that would permit them to detach the Austrian provinces of Lombardy and Venetia. The fighting went unexpectedly poorly for the Austrians, however, convincing them that their initial aims were unrealistic; at the same time, increasing agreement over the probability that Prussia would intervene on the Austrian side (thus shifting the balance of capabilities) convinced the French to abandon the demand for Venetia. After a few months of fighting, therefore, it became possible for the two sides to reach a settlement in which Austria gave up Lombardy but kept Venetia.

This example is far from unique. In the 1939 Nomonhan border war between Japan and the Soviet Union—a little-known but quite intense conflict—the Japanese were quite confident that Soviet logistical problems, command deficiencies resulting from Stalin's purge of the army, and the inherent superiority of Japanese troops would force the Soviets to back down. An intense Soviet offensive in August 1939 demonstrated the inaccuracy of this assessment, and by the middle of September the Japanese were forced to admit that "the enemy had won, and everybody knew it."[71] The Soviets were on the opposite side of the overconfidence ledger later that year, however, when they dramatically over-estimated the ease with which they would be able to force Finland to capitulate in the Russo-Finnish War. While numerical superiority ultimately permitted the Soviets to make territorial gains at Finnish expense, Stalin had to abandon plans to impose a communist government on Finland.[72] In each case, intense battles quickly brought expectations into line and hence brought about a political settlement in a matter of months.

It is by this process, then, that a war driven by the informational mechanism logically will come to an end. The question from the perspective of this project is how long this process is likely to take. In formal models, bargaining typically results in a settlement within a small number of rounds of interaction, but nothing within the model dictates how long a single "round" would last. As a result, formal theorists have disagreed about this question, with Powell arguing that settlement will typically occur quickly, while Smith and Stam argue that leaders who operate on the basis of divergent theories of war may alter their beliefs only slowly, with the implication that the informational mechanism can account for quite extended wars.[73]

Resolving this apparent dilemma requires adding just one additional piece of information. As Blainey observes, overoptimism typically appears not simply in the expectation of victory, but in the expectation that victory will come quickly and at relatively low cost.[74] This observation is unsurprising from a theoretical perspective: as anticipated costs grow, the incentive to resolve a dispute politically likewise increases, meaning that a greater disparity in military expectations is necessary for war to occur, and hence that even substantial variation in expectations about the eventual result frequently will be insufficient to bring about violence. For such optimistic expectations to be reasonable, leaders must have theories about how the war will proceed, which predict for example that the superiority of their armed forces will manifest itself in the initial battles and soon put them in a position to dictate terms, or that the costs imposed by the initial few clashes will break the opponents' resolve and force them to sue for peace. When both sides have such optimistic beliefs, not only will war ultimately belie expectations on at least one side (and quite possibly both), it will do so before costs have mounted unduly. Thus, for example, scholars who have advanced an informational interpretation of World War I have argued that the participants saw war as attractive because they expected to win "by Christmas"; it is hard to imagine how leaders on both sides would have been both expecting an intense, multi-year war and so confident of ultimate victory that fighting was preferable to a negotiated settlement.[75] This observation, however, points to a central weakness of informational explanations for World War I—why did the war continue past Christmas 1914?

Hypothesis 3: Wars driven primarily by the informational mechanism will be limited.

Tests of hypothesis 3 must confront two unfortunate complications. The first is that there is no direct quantitative measure of whether a war was driven by the informational mechanism. Thus while case studies can examine the validity of this hypothesis for a smaller sample of wars, quantitative tests must make use of a more indirect testing strategy. In particular, the statistical tests rely on identifying an empirical relationship that would be expected to hold within the subset of informational conflicts.

The speed of settlement in informational wars depends on two factors: how far apart the two sides' initial expectations were, and how quickly they converge over time as new information is revealed. Of these two factors, the rate of information revelation is the more theoretically appropriate, as it avoids selection effects that arise for variables that act prior to the onset of fighting. The problem for tests based on the divergence in initial beliefs, which must necessarily be measured through proxies, is that leaders take readily observable information into account in deciding whether or not to go to war in the first place.[76] This point is clearest in the context of a specific variable, so to that end it is worthwhile to consider the contention that relative capabilities proxy for the degree of uncertainty in a conflict.[77] According to this argument, when the contending parties in a war are relatively evenly matched, both can believe themselves likely to win, allowing for wide variation in beliefs, whereas when one side is substantially stronger than the other, the range for disagreement is much narrower. This claim, although difficult to evaluate directly, is plausible; the important point is that its effects should be felt not with respect to war duration or severity but with respect to war onset. Thus, if beliefs vary more widely in relatively equal dyads, then those dyads should be more likely to fight, but in those rare cases in which war occurs in unequal dyads it will be happening because the beliefs diverge to an unusually substantial degree. As a result, despite the population-wide variation in uncertainty between equal and unequal dyads, in the limited sample of dyads that go to war the degree of divergence will not differ markedly, meaning that the expected time to settlement also should not differ greatly. This point will apply to any variable that might proxy for the general degree of variation in expectations in a disputing dyad.

The speed at which updating occurs is more tractable, however. Under the informational mechanism, participants update their beliefs in response to events on the battlefield and in the diplomatic arena. Updating occurs as leaders learn that their beliefs about relative capabilities, resolve, strategies, or the intentions of outside actors are incorrect. Measuring the degree of updating about the intentions of third parties is not possible in a statistical context, unfortunately: observed interventions may not be informative if both sides expected them to occur, as for example with the German knowledge that France would inevitably come to Russia's aid in 1914, while noninterventions can be quite informative if one side believed that it was likely to get help, as with the aforementioned Danish disappointment with British neutrality in the Schleswig-Holstein Wars. For capabilities, resolve, and strategies, however, information is revealed as military clashes occur. Competing beliefs about the relative capabilities of different armies and the superiority of generals' strategies are put to the test in battle, while beliefs about relative resolve are confirmed or belied by the opponent's willingness to keep fighting once costs start to be imposed. Thus we can expect that once the fighting begins, updating will quickly ensue.

Moreover, more intense fighting will typically lead to faster updating. Limited clashes reveal less about capabilities, strategy, or resolve than full-scale battles, while a series of battles in a short period of time will be more informative than a single one. In the extreme of a very intense war like the Arab-Israeli Six-Day War, a few days of fighting may suffice to bring the two sides' expectations into alignment. The opposite extreme is provided by a case like the Russo-Polish War of 1919–21, in which the new Polish government and their Bolshevik opponents effectively put their war on hold shortly after it started so that the Bolsheviks could fight the mutually detested Whites in the Russian Civil War.[78] Because fighting between the two sides was extraordinarily limited in the initial phases, both sides could persist in mutually incompatible expectations for some time. These observations thus imply that more intense fighting— more frequent and deadlier battles—will be associated with quicker

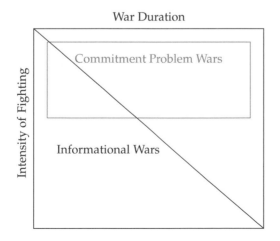

FIGURE 1.2 Informational wars and war duration and intensity

settlement, at least when the informational mechanism is involved. Figure 1.2 illustrates the realm in which wars driven by the informational mechanism will typically be found—intense wars will reach settlement quickly (and hence be short), while less intense wars may be longer but, because they are less intense, are still limited.

> Hypothesis 3a: Wars driven by the informational mechanism will reach settlement more quickly when fighting is more intense.

The identification of a quantitatively tractable hypothesis for the informational mechanism brings us to the second complication for empirical tests, however. While intense informational wars will reach settlement quickly, the same relationship between war intensity and the speed of settlement will not necessarily apply to commitment problem wars. The inclusion of long, intense commitment problem wars in the same empirical sample as informational conflicts thus may disguise the relationship between war intensity and war duration in the informational conflicts. Chapter 2 examines this problem in greater detail and identifies a strategy using the interaction between war intensity and the proxy for the presence of commitment problem wars to more accurately test hypothesis 3a.

Limited Wars II: Misbehaving Leaders

A third logic of war starts from the observation that, because the people are the ones who, through conscription and through taxes, pay the costs of war, they should naturally be disinclined to wage it.[79] If war happens nonetheless, this argument supposes that it may be because the people are not ones who decide whether or not to fight. War may simply be the "sport of kings," or it may be that leaders use war to achieve personal objectives, whether to improve their hold on power or to achieve political goals domestically or abroad that could not be achieved without recourse to war. From this perspective, one might argue that wars, whether limited or unusually large, occur because leaders are attracted by their benefits and simply are not deterred by their costs, which they after all do not have to pay.

On closer analysis, however, I argue that leaders are far less free than this initial perspective implies. All political systems constrain the freedom of leaders to impose preferred decisions unilaterally, whether through institutional checks like the oversight of democratically elected legislatures or practical obstacles like the obstinacy of recalcitrant generals. In the extreme, leaders must worry about being displaced, while publics always have the extreme option of simply refusing to fight. From this perspective, war remains a feasible policy option only so long as the

[42]

leader can convince her constituents that doing so serves their interest. The leader may, of course, resort to all manner of rhetoric and lies in an effort to convince the constituents that launching and continuing the war is worthwhile, but her account of why the war is worth fighting must be reconcilable with publicly known events from the war. To the extent that the leader's accounts of ongoing successes and promises of eventual victory are contradicted by the observation of military stalemate or defeat and the demonstrable falsity of past promises, public and institutional constraints will begin to bind more tightly, until the point at which the leader must either acquiesce to a settlement or be evicted from office and replaced by someone else who is willing to do so.

Ultimately, I argue that the precise path of a war driven by this mechanism is determined by the reason why the leader resorted to war in the first place. In diversionary wars—in which leaders attempt to use war to improve their hold on power—the requirement that the war be politically salient (so as to divert public attention) means that leaders are more constrained in their ability to systematically misrepresent the way that the war is going; as a result, these wars tend to be short. By contrast, when leaders use war to pursue parochial policy objectives such as colonial gains overseas or domestic political programs that would be undermined by defeat abroad, they can afford to limit public attention to the war and hence in at least some cases may be able to systemically misrepresent how well the war is going to a far greater degree. These conflicts, which for shorthand I refer to as "policy wars," thus can last much longer, *but only to the extent that the leader can keep the public distracted*, which in turn implies that the war cannot be particularly intense. Thus I argue that both types of principal-agent wars necessarily will be limited.

Principal-Agent Problems in Domestic Politics

Historians and political scientists have identified a range of domestic political explanations for war. Diversionary war—in which leaders use war either to distract the public from other troubles or to stave off challenges to their hold on power—provides likely the best-known example.[80] Leninists famously argued that the aggression associated with imperialism arose because the state, which represented the interests of the capital-owning minority, sought to address those capitalists' demands for new markets.[81] Other studies have focused on the way in which domestic coalition-building has contributed to external aggression, as with the argument that a coalition of industry and Prussian agriculture led Imperial Germany to a counterproductive policy that alienated both Russia and Britain.[82] Alternately, in some cases domestic

[43]

political pressure has forced reluctant leaders into war, as with the Spanish-American conflict, in which popular pressure for war with Spain, famously encouraged by Hearst's newspapers, ultimately overcame President McKinley's reluctance to fight.[83] Similarly, Fidel Sánchez Hernández, president of El Salvador during the 1969 Football War against Honduras, claimed after the conflict that public pressure for war in response to Honduran mistreatment of Salvadoran nationals was so high that, had he not ordered the invasion when he did, he would have faced a coup within twenty-four hours.[84]

Ultimately, these arguments about domestic politics can be divided into two camps. On the one hand, those in which a reluctant leader is forced by domestic pressure into war ultimately rely on some other mechanism to explain why the public believes war to be in its interest. By contrast, those in which the leader, or an unrepresentative coalition that forms the government, is more war prone than the public can be captured by the logic of the principal-agent problem. This problem was first identified in economics in the relationship between an employer and an employee: at its core, the problem is that the employee—the agent—is being paid to perform certain tasks for the employer—the principal—but has incentives to shirk, reducing effort or engaging in activities of which the principal would not approve, if she can get away with doing so.[85] Shirking in turn is possible because the agent is better informed about her effort and the difficulty of the task that she is undertaking than is the principal. Applied to politics, the principal-agent logic starts from the observation that although effective political rule requires the centralization of power, rulers must always worry about the final sanction of (potentially violent) removal by the public. In this context, we can think of the leader as the agent, empowered by the public to adopt and carry out policies on its behalf. This possibility means that even the most institutionally unconstrained leader must concern herself to some degree with the preferences of her constituents (even if she refuses to think of them in those terms). Thus there are risks to deviating from the preferences of constituents, even at the same time that deviation may be personally rewarding. It is this tension between the potential benefits and costs of hewing to the socially optimal strategy as opposed to deviating for personal gain that lies at the core of principal-agent theory.

When thinking about what goals leaders might have when diverting policy away from socially optimal strategies, the natural starting point is to assume that leaders wish to maintain and strengthen their hold on power.[86] In any system, retaining power carries with it personal benefits that leaders are loath to surrender, while in many political systems loss of power may be associated with exile, loss of freedom, or even death.[87]

[44]

Because regular people care relatively little about the specific identity of their leader in contrast to the quality of that person's leadership, there is a natural divergence of preferences between governor and governed. As the leader, in the course of carrying out her responsibilities, has access to information not available to the regular public, she has an opportunity to shirk, claiming that policy decisions serve the general interest when in fact they are undertaken to serve her interests: for example, by strengthening her hold on power, or by pursuing policy interests that benefit a small coterie of supporters but not society more generally.

Misbehaving Leaders and the Conduct of War

The best-known argument about domestic politics and war duration and severity builds on this intuition. Goemans argues that, while all leaders fear that losing a war will have negative consequences for their hold on power, leaders of partial democracies have reason to be particularly concerned.[88] Leaders of full democracies expect that defeat in war will result in the loss of power, but they are protected by the democratic rule of law from worse consequences. Leaders of autocracies, on the other hand, expect punishment (exile, imprisonment, or death) should they lose power, but their hold on power is generally sufficiently secure to ensure that they need not fear the consequences of a lost war. Thus it is the leaders of partial democracies, where the move toward democratic rule has left leaders less secure in their hold on power without providing them with guarantees of personal well-being once they leave office, who have the greatest incentive to gamble for resurrection, refusing to settle a losing war in the hope that a miracle will restore their fortunes.[89] This argument thus constitutes a potential explanation for long, bloody conflicts: such wars occur when leaders reject settlements that their constituents would accept, if equally informed and empowered to negotiate, because they believe that doing so is in their own interest. More generally, a domestic-political explanation for unusually destructive wars would simply observe that, because they are insulated from the immediate costs of war, leaders may choose to continue wars that their publics would bring to an end.

This argument focuses on the incentives leaders have to implement socially suboptimal policies. The logic of principal-agent problems demonstrates that the agent's (i.e., leader's) incentives are only part of the story: leaders are monitored by society, which can impose constraints on leaders' ability to continue wars against constituents' interests. These constraints come in a number of forms. At one extreme, if a war is sufficiently unpopular (as happened to some degree with the Russians in

[45]

World War I) or a leader is sufficiently despised (as happened in Idi Amin's Uganda during the Uganda-Tanzania War), the population may simply refuse to fight.[90] This option is obviously used rarely, as its adoption effectively grants victory to an opponent that is likely to have preferences that diverge sharply from those of the local population. That said, this possibility is frequently a concern for leaders even when it does not arise. Thus, for example, by the end of World War I the troops of all the major participants demonstrated sufficient unrest to seriously concern their leaders, including the often-forgotten mutinies among French troops at the time of the Nivelle Offensive in spring 1917 on top of the revolutions in Russia and the Kiel mutiny in Germany.[91] Leaders may also find themselves directly constrained by other significant figures or branches of government. Congress's power of the purse makes it possible for the legislature to refuse to fund efforts in an ongoing conflict, in effect forcing the president to find a way to bring a precipitous end to fighting. In practice, these types of constraints too are exercised rarely, as any politician will be extremely chary of taking actions that can be portrayed as beneficial to the "enemy." That said, in both cases the knowledge that such actions can be taken can force leaders to adopt different policies from the ones that they would choose were they entirely unconstrained. To cite a recent example, it is likely that the American surge in Iraq, undertaken in 2007 after significant domestic debate, would have been both quicker and larger had President Bush not had to worry about substantial Democratic opposition in both houses of Congress.

More frequently, leaders may simply be removed from power. Indeed, leaders on the losing side of a war are frequently replaced prior to war termination. War termination is often preceded by a shift in the ruling coalition on one side, as with the two revolutions that preceded Russia's withdrawal from World War I or the shift from the Truman administration to the Eisenhower administration that arguably hastened the end of the Korean War.[92] Even when leaders end a war prior to removal from office, they may do so because they fear that further fighting would only weaken their hold on office. Thus, for example, a significant motivation for Emperor Hirohito's decision to call for an end to Japanese resistance in World War II was his desire to preserve the institution of the emperor and the imperial house.[93] Similarly, in the Persian Gulf War, Saddam Hussein was willing to accept humiliating conditions imposed by the coalition in part because he needed to redirect his efforts to put down the developing revolt against his rule.[94]

A close examination of existing work on domestic politics and war provides reason to think that these constraints are quite active. Quantitative studies have found far more limited evidence for diversionary war than early work expected.[95] In particular, while domestic political factors often are related to the use of force, the relationships typically are not what the diversionary hypothesis would predict.[96] Thus in contrast to the expectation that presidents facing reelection would be more likely to start a foreign conflict to secure their hold on power, presidents have in fact been substantially *less* likely to begin foreign adventures as elections approach.[97] Indeed, in a more general study, Chiozza and Goemans find that conflict initiation is substantially higher when leaders have a secure hold on power, precisely the opposite of what one would expect from the diversionary hypothesis.[98] Similarly, the tremendous boosts in popularity that leaders sometimes gain from crises such as the September 11 attacks on the United States turn out to be less manipulable than initially might have been thought. On average, these gains are quite small, while the big gains in popularity—the conventional "rally 'round the flag" effect—occur only in cases in which the country was subject to an unambiguous external attack of the sort that leaders simply cannot engineer whenever it would be convenient.[99]

One could of course argue that the constraints that appear to limit the ability of leaders to resort to diversionary war are lifted once war begins; after all, it has long been a truism that the nation rallies behind the leader once war has begun, and hawks opposed to settlement repeatedly have used this expectation to accuse moderates of treason.[100] Yet a closer examination reveals reasons for caution here, too. To the extent that the public rallies behind the executive, they often do so because society generally accepts (without substantial misrepresentation by the leadership) that war is in the national interest. Thus, for example, while there existed a substantial isolationist sentiment prior to Pearl Harbor, the Japanese attack genuinely convinced a broad range of Americans that Japan and Germany represented serious threats, and therefore that fighting a long, difficult war to remove those threats was a worthwhile task.[101] Just as large rallies in popular support for the president may occur in response to genuine external threats, we may observe the coincidence of a free executive and a long war in a case in which the people are willing to grant the leader freedom to act precisely because the threat is so obvious that there need be no fear that the executive is extending the war unnecessarily.

More important, the common argument that society inevitably rallies behind the government in war is simply historically incorrect. Indeed, in no recent American war has the US government been free from criticism. Eisenhower won election in 1952 in part by criticizing the Truman administration's policy in the Korean War, while Vietnam of course grew tremendously contentious over time. In the 1991 Persian Gulf War, the congressional resolution to use force passed by only a narrow 52–47 margin, and absent the quick and decisive victory that coalition forces

[47]

achieved the Democrats undoubtedly would have returned to public criticism quite quickly.[102] Republicans repeatedly insinuated that the 1999 Kosovo War was undertaken to deflect attention from the Monica Lewinsky scandal and Clinton's subsequent impeachment.[103] More recently, the 2003 Iraq War was consistently divisive, with John Kerry heavily criticizing Bush administration policy in the 2004 election and a quite public debate over the wisdom of Bush's proposed "surge" in 2007. When presidents hearken back to the bipartisan accord of the World War II period, therefore, they are skipping over half a century in which the occurrence of war hardly guaranteed domestic agreement.[104] Nor is public opposition restricted to the United States. While there almost always are majorities supporting the leader's policies, especially early in wars, minorities in opposition are frequently vocal and sometimes surprisingly effective. Even the paragon of totalitarian control over his country—Adolf Hitler—faced an assassination attempt, which came remarkably close to success, by an opposition that hoped to negotiate a separate peace with the Western powers.

Given the ability of other actors in government, the army, or society in general to reign in misbehaving leaders, the question is when they will choose to do so. If they had perfect control over the leader's activities, principal-agent wars would never occur. In reality, of course, leaders have significant advantages. The first is that imposing constraints on leaders in the context of war is risky and potentially costly: generals and politicians who oppose orders face charges of treason, and soldiers who refuse to fight may be shamed or even executed. People who are uncertain about whether the leader's decisions are reasonable thus face incentives to stay quiet. The second advantage is that leaders are better informed about the true situation in the war that most or all of their potential opponents—they have private reports from the battlefield and know the details of secret negotiations with the enemy.[105] The existence of this information gap is critical. When the principal-agent mechanism is responsible for war, fighting is not in society's interest; leaders therefore can avoid resistance only by lying. So long as the leader can convince her supporters that her publicly professed optimism is justified, or that the opponent is as resistant to settlement as she claims, she will be able to continue the war unconstrained.

Thus, just as in the informational mechanism, the continuation of principal-agent wars over time requires that expectations about the future course of the war not converge: the more that society knows about the true state of affairs, the harder it will be for the leader to avoid the imposition of constraints that force her toward a settlement. As a result, the revelation of information through fighting and negotiations will steadily undercut the leader's narrative about why the war is in the national interest, ultimately bringing about war termination when either the leader is replaced by someone more amenable to peace or is backed into a corner and forced to agree to settle. As in informational conflicts, the faster that information is revealed, the faster that we would expect to see war end.

To summarize, leaders start principal-agent wars because they hope to achieve personal benefits, but they can continue them only so long they can convince society to back them, by maintaining a plausible story about why further fighting will be in the national interest. As the war continues, however, the leader's misrepresentations will become increasingly apparent, as past promises are not achieved and battlefield setbacks make the opponent's terms seem more reasonable. As time passes, therefore, the leader will be increasingly constrained in the positions she can advance; ultimately she will either be forced into an undesired settlement or replaced by a leader who is willing to end the war. This argument thus implies that principal-agent wars will be limited in duration, intensity, or both.

Hypothesis 4: Principal-agent wars will be limited.

Indeed, there are two differentiable types of principal-agent wars, distinguished by the end that the leader is pursuing. The first are diversionary wars, in which leaders see external conflict as a means to improve their hold on power. The second, to which I refer as "policy wars," cover conflicts in which the leader pursues a specific policy interest that the public

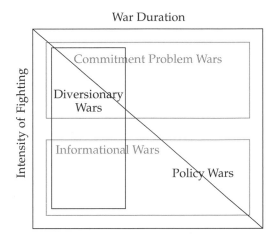

FIGURE 1.3 Principal-agent wars and war duration and intensity

either does not support or would not be willing to pay the costs of war to achieve. While the basic logic of the principal-agent mechanism is the same in each case, the difference in the leader's goals turns out to have important implications for the way in which the conflict is conducted and hence for the nature of the war. Figure 1.3 illustrates my predictions for where each type of principal-agent war will typically fall in the theoretical space, again identifying a basic trade-off between intensity and duration that keeps these conflicts limited.

The point of diversionary wars is to divert attention from the reasons why leaders might have lost popularity or come to face threats to their hold on power. As such, the war must be a public event: the people cannot be distracted by something that they do not know about. More precisely, if the war is intended to distract focus from other matters, it must be more salient than those matters; if it is a forum for demonstrating the leader's capabilities, the public must be able to observe the leader in action. By raising the salience of the conflict, however, the leader loses much of her ability to misrepresent how it is going. As a result, when things go wrong—as they quite often do, given that the leader typically is adopting a risky policy in the hope that it will salvage her position—it will be hard to present the available facts in a good light. Consequently, she will have a hard time making a convincing case that the war should continue, and if she tries she may find her ability to do so effectively quite constrained. The Falklands War, discussed in chapter 7, provides an obvious example of this dynamic. In another case, Idi Amin's invasion of Tanzania in 1978 is frequently attributed to his belief that the army was about to overthrow him and thus that by manufacturing a territorial dispute with Tanzania he might succeed in distracting the army and thus securing his hold on power. In the event, when the Tanzanians launched a counterinvasion, the soldiers of the Ugandan army simply refused to fight, with the result that Amin was rapidly driven from power and the war ended in a matter of months.[106] In short, if a diversionary war starts poorly, the public will quickly realize that the leader has been lying to them; in the less-likely scenario in which it begins well, the leader will usually have achieved her goal and be open to settlement. In either case, diversionary wars will typically be short.

Hypothesis 4a: Diversionary wars will typically be short.

Leaders can also start wars to pursue idiosyncratic personal interests of which society would not approve. Because the goal of these wars is not to influence public opinion, it is not necessary that the public be able to observe them. Indeed, these leaders will benefit from having an uninformed public that is unable to critically evaluate their claims about the

[50]

necessity and course of fighting, as an uninformed public will be more likely to give the leader the benefit of the doubt when she asserts that fighting is in the national interest. As a result, we can expect that leaders in these conflicts will make an effort to limit the availability of information about the war effort, and in particular to limit critical perspectives. Thus, for example, several scholars have argued that Lyndon Johnson escalated the Vietnam War because he believed that allowing South Vietnam to go communist would torpedo his ambitious domestic agenda, headlined by his Great Society program, much as the "loss" of China had hamstrung President Truman before him. Johnson thus authorized a steady expansion of the Vietnam War, consistently misrepresenting increased efforts that were designed simply to keep the situation in Vietnam from completely falling apart as the last pieces necessary to achieve a final victory, while discouraging coverage of events on the ground.[107] This situation could also arise when leaders believe that the public does not accurately perceive the national interest, as for example when Franklin Roosevelt deliberately misled the American public about the extent of the assistance that he was providing to Britain in the months prior to the attack on Pearl Harbor.[108]

There are limits to what leaders can hide from their constituents, of course. Modern media and a free press mean that events anywhere in the world can be reported quickly and in detail, so that no country now can wake up one morning to discover, as Britain repeatedly did, that it has acquired a new colony of sometimes dubious actual value in a distant part of the world.[109] The emergence and spread of this technology thus has undoubtedly limited the ability of leaders to fight wars that their publics would not endorse.[110] There is, however, still substantial variation in the speed with which publics will learn about ongoing wars and the urgency with which constraints on leaders may be imposed. In general, members of the public tend to be better informed and to care more about things that directly affect them or people whom they know.[111] From this perspective, the more salient a war is and the easier that it is to understand and interpret events within it, the quicker the public will develop clear ideas about what is happening and thus the less leeway that leaders will have for continuing wars that their constituents would not endorse.

Salience in turn can come in multiple forms. An increase in death tolls, conflict over highly symbolic or strategically important stakes, and fighting within the territory of the country in question would all tend to increase the salience of a particular conflict, thus hastening the speed at which the public learns and limiting the leader's flexibility. The observation that Americans who know someone who was injured or killed in the 9/11 attacks or in the Iraq War were more likely to disapprove of

President Bush is consistent with this argument.[112] Differences in the speed of learning can also arise from variation in the nature of the war, with the distinction between conventional wars and insurgencies particularly important here, as citizens (and even governments) have to identify alternate methods for assessing how well the war is going.[113] Whereas in conventional wars the participants generally fight distinct battles that provide an obvious focal point for attention, in insurgencies such battles are far rarer. From this perspective, the quintessential low-salience war would be a low-intensity guerrilla conflict in a peripheral part of the world fought over stakes of little importance to regular people. These sorts of principal-agent wars thus will be largely the prerogative of great powers, which have the wherewithal to undertake military interventions in far-off lands even over issues of relatively low immediate importance.

Hypothesis 4b: Policy wars may be of extended duration, but only if they are not particularly salient.

Case studies, especially in chapter 7, shed light on several of these hypotheses. As with the informational mechanism, however, statistical analysis is complicated by the lack of any *ex ante* quantitative indicator of whether a war was driven by the principal-agent mechanism, which forces me to resort to indirect tests. Two additional testable hypotheses follow from the theoretical discussion above. The first, already discussed, is that quicker information revelation should be associated with quicker settlement. The second is that wars will tend to be more limited the more tightly leaders are constrained. Constraints come from several different sources, including the army, other centers of power within the government, and at an extreme the citizenry as a whole. In practice, most of these constraints (such as the presence of additional centers of power) are more likely to exist, or are likely to take on a more substantial form, in democracies.[114] Some of course exist in every political system: soldiers can always refuse to obey orders, and people can always engage in strikes, large protests, or revolutions. However, even in these cases, democratic leaders will have fewer options in responding to such actions: an autocrat has the option of killing dissidents en masse and without trial, something a democrat will have much more trouble doing. Thus one prediction of this model is that increased democracy will result in reduced opportunities to pursue personal gains at public expense. Most obviously, this argument implies that democratic leaders will have greater difficulty in refusing to settle an unsuccessful war, meaning that wars in which the loser is more democratic will tend to be more limited. In addition, a number of scholars have argued that democratic leaders, anticipating greater constraints, are more cautious when initiating wars, which should imply that the wars that they initiate will be more limited.[115]

> Hypothesis 4c: Principal-agent wars will end more quickly when fighting is more intense.

> Hypothesis 4d: More democratic war losers and war initiators will be associated with quicker settlement.

What separates the few unusually destructive wars from the many that either end more quickly or are fought less intensely? This chapter argues that the most destructive wars are typically driven by commitment problems. Specifically, I argue that preventive wars driven by situational commitment problems—in which a power anticipating relative decline starts a war to stop the decline from occurring—are unusually difficult to resolve, because the drive to prevent the decline entails large war aims that the declining power is reluctant to reduce, even in the face of military difficulties. In addition, I identify a novel mechanism, which I refer to as the dispositional commitment problem, in which targets of preventive wars attribute aggression not to the initiator's fear of decline but to an aggressive disposition, with the implication that the initiator will continue to launch unprovoked attacks if given the opportunity. Given these beliefs, no negotiated settlement can be expected to last, and thus it makes sense to pursue the opponent's unconditional surrender.

This chapter also examines in detail two other mechanisms that can produce wars but that for related reasons do not produce unlimited wars. In the informational mechanism, wars begin because leaders have divergent expectations about how fighting will go. Once fighting starts, those expectations begin to converge, more quickly when fighting is more intense, until they are close enough that settlement becomes possible. In principal-agent conflicts, leaders start wars that serve their own interests, but not those of their constituents. These wars are sustainable, however, only to the extent that the leader can hold onto power and avoid pressure to settle. Constraints on leaders in turn become stronger as their constituents realize, again on the basis of events on the battlefield and at the negotiating table, that they have been misrepresenting how well the war can be expected to go. This argument thus implies that both informational and principal-agent wars can be either long or intense, but not both.

[2]

Research Strategy and Statistical Tests

In chapter 1 I made general arguments about the sources of limited and of unusually destructive wars and more specific predictions about the implications of different logics of war. This chapter examines those predictions quantitatively, focusing first on general predictions about the sources of unusually destructive wars and then turning to specific hypotheses drawn from the different mechanisms. Before doing so, however, I will introduce and justify the multiple-method research strategy used in this book, which is particularly appropriate for a project like this one, which combines multiple theoretical arguments—complicating pure case analysis—with hypotheses that at times cannot be tested quantitatively. After that discussion, I turn to the data used in the quantitative analysis focusing on measures of war destructiveness and my operationalization of key independent variables. The subsequent analysis begins with tests of general predictions about war destructiveness before more briefly addressing subsidiary hypotheses.

RESEARCH STRATEGY

The existence of multiple mechanisms that all provide a sufficient explanation for war, but that differ in their implications for the course and destructiveness of the wars that they cause, raises a number of significant challenges for empirical research. On the one hand, equifinality raises the prospect that different cases may be driven by diverse causal processes, thus potentially casting doubt on the results of purely qualitative studies, especially studies of a relatively small number of cases. In short, we cannot count on findings from a case driven by one mechanism to apply to a case driven by another. This problem provides a good reason, beyond the many that already exist, for quantitative tests of the hypotheses generated in the past two chapters. Moreover, quantitative approaches provide a useful check on the tendency of researchers to extrapolate from the most familiar cases, a particularly significant problem if the most familiar cases are unrepresentative, as with the tendency of international relations researchers to know more about particularly deadly or consequential wars.

At the same time, central variables in the bargaining model are difficult to observe and to operationalize, casting doubt on the results of purely quantitative studies. Because no available variables can reliably identify which mechanism is driving conflict in a particular case, I am forced to rely on proxy variables or on more indirect tests of central hypotheses in quantitative work. These tests are still quite useful, but there is always the concern that indirect tests may pick up relationships that exist for reasons other than those identified by theory and that proxy variables may not be operating in the manner expected. Moreover, several arguments in this book simply cannot be tested quantitatively, either because (as in the case of unconditional surrender) the number of relevant historical cases is too small or (as in the case of predictions about the size of war aims) there is no good way to quantify a variable of interest validly and reliably. Case studies thus provide important additional tests of my arguments. In addition to their obvious utility in testing broad predictions about unconditional surrender that cannot be tested statistically, case studies are particularly useful when there is potential for disagreement about *why* an observed empirical relationship exists—the only way to adjudicate between competing explanations for the same phenomenon is to derive and test additional, divergent implications of the competing theories. Even if such tests can only be conducted in a limited number of cases, they can still provide a substantial increase in confidence in the validity of one interpretation relative to another. For this reason, the approach taken here is to use a multiple-method research strategy that combines statistical analysis with close case studies. The statistical analysis provides confidence that results hold across a wide range of wars through history; the case studies provide confidence that broad patterns found in the statistical results are present for the reason that theory predicts that they should be present.

With respect to the qualitative analysis, the next question is what cases a study should examine. Given the need for significant variation on the dependent variable, I must examine both large and small wars. I also wish to ensure that results do not simply reflect the bias of the field toward examining well-known cases—in other words, if the well-known cases (for example, the major-power wars) result from a different causal

[55]

process, I want to vary my case selection to be able to identify those differences.[1] At the same time, I also want to ensure that readers have an opportunity to check my interpretations against their own readings of cases with which they are familiar.

Given these disparate goals, I ultimately opted for a bipartite case-selection strategy. The first approach was to select three cases from a stratified random sample, choosing one each from the sample of short wars; long but low-intensity wars; and long, high-intensity wars.[2] This selection process identified the Anglo-Iranian War of 1856–57 (short), the Franco-Turkish War of 1919–21 (long but low-intensity), and the Paraguayan War of 1864–70 (long and high-intensity). These case studies force me to address unfamiliar conflicts, thus helping to allay concerns that cases were chosen because they were known to conform to theoretical expectations or that intentional case selection (which often focuses on more familiar conflicts) unintentionally introduces bias into the analysis. The drawback to random case selection, however, is that it often highlights cases that are unfamiliar to readers—meaning that they must put greater trust in the researcher's honesty in presenting relevant information—and for which evidence is often scarcer, with the result that tests of core hypotheses may be less definitive.

Thus I supplemented the randomly selected cases with a number of cases chosen intentionally as conflicts that would be more familiar to readers and that might be seen as providing a particularly fair test of key arguments. The Persian Gulf War of 1991 was a short conflict that provides a useful test of hypotheses about short wars. The Falklands War of 1982 is frequently seen as the preeminent example of a diversionary war and thus sheds important light on arguments about the logic of principal-agent problems in domestic politics. The European theater of World War II constitutes by most measures the most destructive war in history; any serious attempt to explain such wars must grapple with this case. I examine all three of these cases in significant detail in later chapters. In addition, given the significance of case study tests for my argument about unconditional surrender (which I cannot test quantitatively), I present minicases of the Crimean War, the Pacific component of World War II, and the Iran-Iraq War in chapter 5. Although these cases receive a more superficial discussion than others in this project, their inclusion provides an opportunity to further highlight the ways in which the commitment problem mechanisms act, as all involve significant preventive motivations for war, while two of the three (all but the Crimean) involve one side sincerely refusing to negotiate, in line with the unconditional surrender mechanism.

With case selection complete, the remaining question is what sort of evidence I will use to test the various hypotheses. The most important

evidence concerns what leaders do: lowering political demands or raising them in response to battlefield and diplomatic events, adopting riskier or less risky military strategies, expanding wars to new participants
or keeping them limited, and so forth. After all, these are the decisions
that determine the length and severity of wars. At the same time, however, specific decisions frequently are open to multiple interpretations,
some of which would be compatible with hypotheses advanced here and
some of which would not. Distinguishing among different interpretations requires that I examine the beliefs and statements of leaders. These
statements introduce additional challenges, in that leaders may dissem-
ble or be inconsistent in their beliefs. Given these concerns, I privilege
private statements, which are less likely than public discourse to be rhetorical, and attempt to verify that stated beliefs and rationales are consistent across multiple data points. I also ask whether stated beliefs are
consistent with observed behavior: if, for instance, a leader professes renewed optimism after a battle but privately lowers her political demands, that observation provides reason to think that the stated optimism
is insincere.

Given the number of cases, the individual studies rely primarily on
secondary sources, although I do make use of primary sources for cases
for which they are readily available, typically in conflicts involving Western democracies. Unsurprisingly, there is a far more extensive literature
on some of these cases than on others, and thus evidence is more definitive for some arguments than for others. Given these limitations, I indicate when data limitations reduce certainty in conclusions. Where
possible, this problem was also addressed by the use of French, German,
and Russian-language sources.

QUANTITATIVE ANALYSIS

The remainder of this chapter introduces the quantitative data used in
the statistical analysis, discusses statistical specifications, and presents
and discusses statistical results. There are several central findings. First,
and most important, wars that are preceded by significant shifts in relative capabilities, which I argue should be associated with fear of future
decline for one side in the war and hence with the commitment problem
mechanism, tend to be substantially longer, deadlier, and more economically costly than wars in which such shifts are absent. This result, which
is statistically robust, provides strong support for the argument that the
commitment problem mechanism is responsible for the most destructive
wars. Moreover, further analysis finds that the measure of commitment
problems is associated with greater difficulty settling wars but also with

[57]

quicker conquest, as well as with a higher probability that leaders adopt relatively risky military strategies and that wars end through conquest rather than settlement. All of these findings are consistent with the theoretical argument that leaders launching preventive wars will typically resist settlement absent a major victory, but also will adopt risky military and diplomatic strategies that increase the probability and speed of decisive military victory or defeat.

In addition, the statistical analysis finds that, once we set aside the sample of commitment problem wars (which tend to be both long and intense), increased war intensity is associated with quicker settlement. This finding is consistent with the theoretical prediction that wars driven by the informational or principal-agent mechanisms are necessarily limited in either intensity or duration. These mechanisms, then, cannot account for the most destructive wars, which are both long and intense. In addition, the statistical results reveal that greater institutional constraints on leaders tend to be associated with quicker settlement, consistent with the argument that whatever incentives leaders may have under the principal-agent mechanism to extend wars unnecessarily will eventually be overridden by the constraints that they face.

Data and Measurement

Following the standard approach in international relations, the universe of cases for this study consists of all interstate wars in the post-Napoleonic period, where an interstate war is defined as a violent clash between two internationally recognized states in which at least a thousand battle-related fatalities occur.[3] The starting point for this list is version 4.0 of the Correlates of War (COW) list of interstate wars. Following the standard approach in quantitative studies of the duration and outcome of war, I disaggregate several large multilateral conflicts like World War II.[4] The specific coding rule for disaggregation stipulates that no two countries may be principal belligerents in multiple wars at the same time and requires that the political issue at stake in the fighting differ substantially across fronts. This approach produces a total of 103 primary observations.

In a number of cases, however, reasonable doubt exists as to whether or how a specific conflict should be included in the dataset; to facilitate robustness checks, I include all questionable cases but flag them to permit their inclusion or exclusion as necessary. There are five different reasons why observations may be flagged. In some cases, particularly in a number of nineteenth-century Latin American conflicts, reasonable doubt exists as to whether the participants suffered more than a thousand deaths. Others, as with the Israel-Syria clash in Lebanon in 1982 or the American intervention in Afghanistan in 2001, arguably are better described as internationalized civil wars. In a few cases, the political leadership on one side opposes and tries to prevent the fighting, as with the Moroccan sultan's opposition to conflict carried out by Rif tribesmen against the Spanish in 1909; this situation is potentially problematic given the theoretical assumption that leaders are the ones choosing to go to war.[5] Fourth, given past disagreement about when and how to disaggregate multilateral wars, I also include observations that reaggregate the separate conflicts. Finally, in some cases fighting stops for a short period before resuming, raising potential doubts about whether the case should be considered a single case or two separate wars. I thus include observations that aggregate or disaggregate wars over time differently. Supplementary files available online provide the data used in the analysis, a full discussion of the universe of cases and justification for specific coding decisions, instructions (a Stata .do file) to replicate the main findings, and an appendix containing results from statistical robustness checks.[6]

Measuring the Destructiveness of War

The suffering imposed by war comes in many forms, most obviously dead and wounded soldiers, but also civilians who are harmed by fighting, those who suffer and die from disease and privation associated with war, as well as those who suffer the lesser but still real opportunity costs that arise when the state diverts money to pay for war. No one measure can capture all these costs, and various possibilities involve different trade-offs. For the analysis here, I ultimately use three measures: total battle deaths, war duration, and government spending, each with its own advantages and disadvantages.

The most obvious measure of the cost of war is how many people are killed. Death data pose both conceptual and data quality challenges, however. Conceptually, the responsibility of war for deaths can occur along a continuum, ranging from clear cases such a soldier who is shot on the battlefield to very indirect incidents such as an elderly woman who succumbs to pneumonia in part because war restrictions have left her less well nourished.[7] Contrast, for example, three deadly helicopter crashes, one while the pilot maneuvers under fire, one on returning from a mission, and one during training far from the battlefield: on the one hand, none might have occurred absent the war (although training accidents certainly happen in peacetime), but all are to some extent accidents and hence less unambiguously war deaths than the case of a soldier felled by an enemy's bullet. In part for these reasons, and in part because in war records may be destroyed, falsified, or simply not kept, precise records of deaths in war are frequently simply unavailable, especially for incidents less directly connected to the battlefield. Given the data quality concerns, this study focuses on deaths in battle. Because of concerns about existing sources on battle deaths, I collected this data myself, relying primarily on Clodfelter's encyclopedia of war statistics.[8]

The duration of war presents a different trade-off. Relative to other measures, war duration is particularly reliable—it is much easier to know when fighting started and stopped than to identify exactly how many soldiers died.[9] The existence of war imposes costs on society that continue to be borne as long as fighting continues: all else equal, a longer war is a more destructive one. Of course, in reality we cannot assume that all else is equal—from the perspective of the United States, the Vietnam War was clearly longer than World War II, but World War II was undeniably costlier. Thus with war duration we are confident in the reliability of our measure of the destructiveness of war, but its validity is somewhat reduced. As with battle deaths, to maximize the validity of the war duration measure, I rely on novel data, in this case dates of war onset and termination collected for the War Initiation and Termination project.[10] This data focuses specifically on the date on which fighting starts and stops, rather than, as in the Correlates of War (COW) dataset, diplomatic developments like declarations of war or peace treaties.[11] As with war deaths, substituting the standard COW data for the measure used in primary analyses results in effectively identical results.

The economic cost of a war is a third way of capturing its destructiveness: money spent to pay soldiers and buy weapons cannot be spent on food or to hire teachers. As with deaths, precisely measuring the cost of a war is complicated conceptually and in terms of data availability. In addition to paying for procurement costs during the fighting, war participants typically must rebuild arms stocks after the war, pay for veterans' benefits, and pay interest on any debt incurred during the war. Moreover, the economy of a country at war suffers from the economically inefficient allocation of resources—farmers who are fighting cannot sow and harvest crops, for example.[12] A reliable measure for a wide range of wars necessarily must focus on a more restrictive set of costs, specifically procurement costs during the conflict. Here I rely on information gathered by Cappella, who collected secondary source data on government expenditures for war as part of a project on how states finance wars.[13] The specific measure uses total spending in the war as a share of GDP in the year of war onset, with GDP data coming from Maddison.[14] Unlike the data I have collected for war deaths and war duration, I have data on monetary cost for only a subset of observations (sixty-four countries in thirty different wars, with shorter and less deadly wars disproportionately unlikely to have data). Results from analyses of

[60]

spending are thus less definitive than are those for other dependent variables, but they do shed some additional light on costs.

Explanatory Variables

Fear of Future Decline

Hypothesis 1 predicts that fear of decline will be associated with more destructive wars. To test this claim, I need some measure of the degree to which participants in a war might fear relative decline.[15] The ideal measure would capture projected future shifts in capabilities. Observed shifts after the start of the war are clearly inappropriate, however, as the conduct and outcome of war can dramatically influence relative capabilities in a way that cannot be anticipated precisely at the outset. Indeed, to the extent that a strategy of preventive war is effective and the rising power is defeated, this measure might indicate that it was in fact the *rising* power (if anyone) that anticipated a decline. Instead, I argue that shifts in capabilities prior to the start of war provide a reasonable, if imperfect, proxy for anticipated shifts in future capabilities. In general, the best prediction that actors have for future developments is a function of current trends: the expectation today that China will continue to rise in the future is based largely on the observation that it has experienced significant growth in the recent past. This tendency can perhaps be demonstrated most convincingly in cases in which extrapolation from current trends proved to be incorrect. Thus, for example, a 1990 survey found that 60 percent of the American general public and 63 percent of leaders viewed economic competition with Japan as a "critical threat," and in the early 1990s Americans were concerned by books that augured increased future competition with Japan, including a prediction of a coming war; these fears dropped off, however, as Japan descended into its lost decade.[16]

Indeed, this approach can capture not only the gradual but apparently inexorable rise of countries like Japan or China, but also windows of opportunity associated with dramatic but temporary drops in relative capabilities, generally resulting from domestic upheaval. Salient examples here—both driven by purges of the military—include the Soviet Union in the later 1930s and Iran following the 1979 revolution. In such cases, we would typically expect that countries that experience decline will recover in the future, with the size of the recovery correlated to the size of the initial decline. The existence of contrasting possibilities does indicate that we need to look at the details of cases before deciding which side faced motivations for preventive war in situations of significant capability shifts. When the shift is associated with an inexorable rise (the United States compared to China today), then the side that has become weaker at the end point fears continued decline. When the shift is temporary

[61]

(Iraq compared to Iran), then the side that is *stronger* might consider preventive war, as its advantage cannot be expected to last. While this approach does not directly measure anticipated shifts, I argue that it provides a useful proxy: as the size of shifts prior to war increases, it is more likely that leaders on at least one side will be acting out of preventive motivations, and those motivations will likely be more severe.[17]

I develop the specific variable used in the statistical analyses using COW's National Military Capabilities dataset, which generates a yearly index of state power based on the size of the military, military expenditures, total population, urban population, iron/steel production, and energy consumption.[18] The degree of shift in capabilities for a pair of countries $\{A,B\}$ at times $t \in \{1,2\}$ can be captured by the expression,

$$\frac{cap_{A_1}}{cap_{B_1}} = k\frac{cap_{A_2}}{cap_{B_2}}, \text{ or } k = \frac{cap_{A_1} * cap_{B_2}}{cap_{A_2} * cap_{B_1}}$$

To ensure that the variable is well behaved I construct it so that $k \in \{0,1\}$, which corresponds to assigning A to be the power that has experienced decline in the period prior to war. For ease of interpretation, I then subtract this variable from 1 to ensure that it is rising in the degree of the power shift. The primary power shift variable used in the analysis uses data for all actors in the dispute; I also construct alternate measures that restrict the variable either to the primary dyad in the war, where the members of the primary dyad are the countries that most historians describe as the most important actors on each side of the war, or to those actors initially involved in the conflict, where initial involvement includes all countries that were actively involved within the first month of fighting. I used time lags of five and ten years prior to the war based on the expectation that leaders would need several years to infer a shift in capabilities; the primary analyses use lags of ten years.[19] To avoid potential concerns about shifts arising as a consequence of war, capabilities data for the year of war onset is the lagged value from the previous year.

This variable is of course an imperfect measure of fears of future decline. Case studies in chapters 3–5 present evidence of significant fear of decline in a number of wars from history. Of these, most score quite high on this measure. Thus, for example, the Paraguayan War, the World War II clash between Germany and the Soviet Union, and the Iran-Iraq War all lie above the seventy-fifth percentile in the size of the prewar power shift. That said, there do exist cases in which fears of decline are not picked up by the measure. The British fear of decline in the Crimean War, for example, hinged on a potential development (the Russian acquisition of Constantinople) that had not happened and thus could not affect

[62]

prewar capability levels, with the result that the capability shift variable is low in this case. On the other hand, a few cases such as the Second Central American War or the Vietnamese-Cambodian War score quite high on the measure despite the absence of any clear evidence that participants worried about future decline. These sorts of cases introduce measurement error, which necessarily reduces confidence in the results, thus providing an additional rationale for qualitative work. That said, the close relationship in the remaining cases provides reason to believe that the fit is not unacceptably poor. Moreover, this proxy benefits from the absence of any other reason to believe that prewar shifts in capabilities would be associated with worse wars.

Revelation of Information

Hypotheses 3a and 4c predict that more intense fighting will reveal novel information to relevant actors more quickly, producing quicker belief updating and hence quicker settlement, at least in wars that are not driven by the commitment problem mechanism.[20] I measure war intensity—defined as the rate at which battle deaths accumulate—using the war deaths data discussed above. When it comes to measuring intensity, of course, we must account for the size of the combatants: a battle in which a thousand soldiers die may be a minor skirmish between Germany and Russia or an epic clash between Guatemala and El Salvador; leaders would obviously learn much more about their relative strength from the latter battle than from the former. The specific variables are thus adjusted for population size, using the measure of national populations from COW's National Military Capabilities dataset. The resulting variable is then logged to correct for skew, to limit the possibility that otherwise outlying observations might drive the results.

Constraints on the Executive

Hypothesis 4d predicts that increased constraints on the executive, either for the war initiator or for the side that is losing militarily, will be associated with quicker settlement. The best available measure of such constraints is the country's level of democracy. The standard measure of democracy in international relations scholarship comes from the Polity dataset, which is the only available measure that provides coverage over the entire post-Napoleonic period.[21] Countries are coded on two composite scales, one capturing the level of autocracy and one capturing the level of democracy; these are frequently combined to generate a single, twenty-one-point scale. Both the autocracy and the democracy scores are generated from several separate variables that capture aspects of democracy like the openness and competitiveness of executive

[63]

recruitment (i.e., how wide a circle of people has a chance to become the leader? how are leaders selected?). It thus is focused primarily on the institutions of government rather than on things like civil society or democratic norms. While this focus may be a problem in many contexts, it is ideal for a study such as this one that is concerned with the way in which domestic political institutions may affect leaders' incentives and constraints.

I generate two different regime variables from the Polity data. The first is the Polity score of the side that loses militarily, with the military loser identified by the War Initiation and Termination (WIT) dataset.[22] Complications for this coding strategy arise in the context of multilateral wars and of wars that end in military draws. In multilateral wars, there is the possibility of an alliance of highly dissimilar regimes, with cooperation between Russia and the Western powers in both World Wars providing the most obvious examples. In this context, averaging scores would be quite misleading, while coding one extreme or the other would be open to challenge. One step here is to limit the focus to those countries that actually could influence central decisions about the war, omitting those (like Poland in the Iraq War) that contribute troops but have minimal say over military and political strategy. This decision is taken for obvious theoretical reasons—the domestic political incentives facing leaders of countries that do not have much sway over whether the war should be settled or continued logically should not much influence the nature of the war—but it has the ancillary benefit of eliminating some cases of allied but dissimilar losers, such as Germany and Finland in World War II. Separately, and extremely conveniently, it happens that in most multilateral wars a single country loses to a coalition, and furthermore when a coalition loses the countries involved generally share similar governance structures. As a result, substantial deviations are quite rare, and turn out not to affect statistical results.[23]

The second regime variable captures whether or not the war initiator was democratic, drawing on the argument that democratic leaders tend to be particularly selective in the wars that they start, initiating conflicts that they can win easily.[24] This variable is a dummy that is coded 1 when the initiator has a Polity score of 7 or higher on the standard -10 to 10 scale.[25] Data for the war initiator comes from the Correlates of War interstate wars dataset.

Control Variables

Finally, I control for a range of variables that other studies have identified as significant predictors of war duration and severity. Given the possibility that rough terrain will slow down the process of fighting and thereby extend wars, I control for the nature of terrain, using data from a prior study by Branislav Slantchev.[26] Contiguity similarly has a plausible association with war duration, as more distant opponents will have greater difficulty bringing force to bear against each other. Contiguity data comes from version 3.0 of the Correlates of War's contiguity dataset; the dichotomous variable captures whether the primary participants in the war shared a land border.[27] Similarly, given the argument that the involvement of additional states in a conflict increases the number of veto players who might reject a possible settlement, I control for the number of major participants in a war.[28] Data on major participants comes from the WIT project, which identifies a major participant as a country that has a substantial influence over its side's political or military decisions.

Several authors have argued that relative capabilities also influence war duration and severity, typically with the prediction that relative equality will be associated with worse wars.[29] Data for relative capabilities comes again from the Correlates of War National Military Capabilities dataset. The relative capability variable is equal to $\dfrac{cap_a}{cap_a + cap_b}$, where cap_i is the sum of capabilities for the primary participants on one side and $cap_a \geq cap_b$. The variable thus ranges between 0.5 and 1, with higher values representing a more unequal distribution of capabilities.[30] I also control for major power involvement with a dichotomous variable that captures major power involvement on both sides of the war, using Levy's identification of major powers over time.[31]

Others might expect that cultural or ideological differences would produce worse wars.[32] I code cultural difference using Huntington's delineation of the major world civilizations around which he predicted that conflict would occur, generating a dichotomous variable that captures whether the primary dyad in a war (coded as above) belong to different civilizations.[33] For political ideology, I generated two related variables. One is based on a division of regime ideologies into democratic, monarchic, communist, fascist, and "other authoritarian" categories; the second subdivides the "other authoritarian" category to allow for progressive/liberal and religious/conservative categories in the nineteenth and early twentieth centuries.[34] In both cases, I code a clash of ideologies if the members of the primary dyad adhere to different regime philosophies.[35]

A final control variable captures each side's choice of military strategy (among the options of maneuver/blitzkrieg, attrition, and punishment/guerrilla war), which has been found in past studies, particularly by Bennett and Stam, to significantly influence war duration.[36] I use Bennett and Stam's codings where possible, while coding observations that do not appear in their dataset myself. The specific variable codes strategy along a single dimension, with the midpoint occupied by a baseline category of conventional attrition, which contrasts to maneuver/blitzkrieg strategies (lower values) and guerrilla/punishment strategies (higher values). Higher military strategy scores thus should generally be associated with longer (but not necessarily deadlier) wars. Inclusion of strategy variables in my analysis is potentially problematic given the argument that the strategy may be endogenous to the mechanism the led to war; in practice, however, results for other variables are unchanged by the inclusion of the strategy variable. I also use the strategy variable to test hypothesis 1b, which predicts that preventive wars will tend to feature the use of riskier military strategies.

Conquest and Settlement

Finally, testing several of the subsidiary hypotheses requires that I have a measure of how war ended, distinguishing between logical possibilities of settlement and conquest. Game theorists have found that generating a compelling model of the conduct of war requires allowing for the possibility that war ends either through a negotiated agreement between the two sides or the military collapse of one side that renders its leaders incapable of continuing to contest control over the political issue at stake.[37] From this perspective, we can contrast the Korean War, which ended with both sides capable of continuing to fight, to World War II in Europe, in which Hitler and leading Nazis would have liked to continue the struggle in May 1945 but lacked the means to do so. The theoretical discussion of commitment problem wars predicts that leaders who fear decline will resist settlement but also will adopt risky military strategies that at times will result in relatively quick conquest; jointly, reluctance to settle and the use of risky military strategies should make conquest more likely in these conflicts. Testing these offsetting predictions requires that I be able to distinguish between types of war termination.

As no data on the nature of war termination in interstate wars exist, I collected this information myself.[38] Conquest occurs when one side loses the ability to sustain organized resistance, which typically forces capitulation to the conqueror's demands.[39] Any form of war termination that falls short of conquest is coded as settlement. Settlement thus covers cases such as the Korean War in which extensive negotiations preceded a signed armistice as well as wars such as the Sino-Indian conflict of 1962, in which India tacitly accepted China's declaration that the war was over by refraining from continuing to contest the disputed border. In other words, as long as both sides retain the capacity to fight, I code the war as

[66]

ending through settlement; if one side loses that ability, as defined in specific coding rules, the war ends through conquest.

The basic approach is to code a war as ending through conquest if the military on the losing side ceases to exist as a viable fighting force (implying that war termination is, from the perspective of the political leadership, involuntary).[40] This can happen, as with Germany in World War II, when all individual units in the army are either destroyed or forced to capitulate, or it can happen, as for example in the Uganda-Tanzania War, when the army on one side simply stops fighting and goes home, leaving the leaders on that side with no way of continuing the war. As with any coding exercise in international politics, however, there inevitably are marginal cases over which reasonable people could disagree. Thus, for example, how should we handle a case like the end of the Pacific War in World War II, in which the Japanese military surrendered in its entirety, thereby losing the capacity to continue to fight, in return for one political concession (the retention of the emperor in a figurehead role) on which the Allies could subsequently have reneged? Alternately, how should we handle a case like the Falklands War, in which both sides retain an ability to fight, but one has been forced out of the territory under dispute and is

TABLE 2.1 Descriptive statistics for dependent and independent variables

Variable	Mean	Median	Standard deviation
War duration (days)	419	120	741
Battle deaths	242,699	8,600	1,093,075
Government spending	0.55	0.11	1.9
Conquest (primary coding)	28 of 103 wars end in conquest		
Conquest (restrictive coding)	21 of 103 wars end in conquest		
Conquest (expansive coding)	35 of 103 wars end in conquest		
Prewar capability shift	0.28	0.24	0.22
Loser's political regime	–3.3	–5	5.6
Democratic initiation	21 of 103 wars were initiated by a democracy		
Log (per capita war intensity)	–6.6	–6.4	2.0
Terrain	0.62	0.6	0.24
Contiguity	61 of 103 wars involve directly contiguous countries		
Relative capabilities	0.79	0.81	0.14
Number of war participants	2.4	2	0.76
Major power war	14 of 103 wars involve a major power on each side		
Military strategy	4.0	4	1.1
Total population (100,000,000s)	0.19	0.067	0.27
Total GDP of Allies	0.40	0	0.83
Own GDP	0.27	0.058	0.82
Cultural/civilizational clash	62 of 103 wars involve a clash of civilizations		
Ideological clash	63 of 103 wars involve a clash of ideologies		

Note: All reported statistics are for the default set of wars used in the primary analyses.

incapable of returning? Because of the existence of such marginal cases, I ultimately developed three versions of the conquest/settlement variable. Under the primary coding, conquest is coded whenever the army surrenders or ceases to exist or when further resistance is impossible (for example because the entire territory of the defeated country is occupied). Under a more restrictive coding, cases initially coded as conquest are switched to settlement whenever the loser's surrender was purchased with minimal and unenforceable concessions, as with Japan's surrender in World War II. Under a more expansive coding, cases are coded as conquest if the war ended in conquest under the primary coding rule or if the war ended when one side overran a geographically separated and discrete region that constituted the central issue at stake in the war, as in the Falklands case.

STATISTICAL ANALYSIS

The first section of the statistical results examines broad hypotheses about the relationship between different mechanisms and the destructiveness of war, focusing in particular on the prediction that commitment problem wars will be unusually destructive. The next section examines several subsidiary hypotheses that, while less directly connected to the overall destructiveness of wars, allow for additional tests of the different mechanisms. The first set of results is thus critical in demonstrating that the broad theoretical predictions drawn from the different mechanisms are consistent with the historical record; the second set provides additional checks that help to increase confidence that the observed relationship exists for the reason that theory posits.

The analysis here makes use of events history models, which are designed to deal with questions related to duration.[41] In an event history framework, observations (here wars) are compared with respect to how long it takes for them to exit the sample, in this case by ending. This approach is obviously appropriate for analysis of war duration, but it also makes sense for examining deaths and spending. Just as each additional day a war continues is a further day of suffering, so each additional death or dollar spent is a further cost that could have been avoided by a quicker settlement. The first set of results use the semiparametric Cox specification.[42] For the second set of regressions, I distinguish between different types of settlement to allow for the possibility that a variable might simultaneously delay settlement and hasten conquest. For these analyses, I use a competing risks regression framework.[43] The final set of tests examines dichotomous dependent variables (whether or not participants resorted to relatively risky maneuver strategies, and whether wars ended

[68]

through conquest or settlement), for which a probit specification is statistically appropriate.

Tests of General Predictions

The single most important hypothesis in this book is that wars driven by commitment problems will tend to be more destructive than those driven by other mechanisms. Table 2.2 presents results for regressions examining the determinants of war duration (models 1–2), total battle deaths in war (models 3–4), and government spending (models 5–6). The reported estimates are variable coefficients (rather than hazard ratios), and as such have a somewhat unintuitive interpretation: a positive coefficient implies that an increase in the variable in question is associated

TABLE 2.2 Predictors of war duration, battle deaths, and government spending

	War duration		Battle deaths		Government spending	
	Model 1	Model 2	Model 3	Model 4	Model 5	Model 6
Capability shift	−1.29*	−1.58**	−1.40**	−1.53*	−2.90*	−2.68*
	(0.53)	(0.53)	(0.52)	(0.60)	(1.37)	(1.21)
Log (war intensity)		0.23*		−0.46**		−0.36
		(0.11)		(0.13)		(0.33)
Democratic initiator		0.71*		0.54		0.30
		(0.33)		(0.38)		(0.58)
Loser regime type		0.044*		0.049*		
		(0.021)		(0.022)		
Terrain	−2.51**	−2.42**	−1.45**	−1.52*	−0.24	−0.80
	(0.52)	(0.60)	(0.54)	(0.71)	(1.52)	(1.64)
Contiguity	0.027	−0.038	−0.24	0.14	0.29	0.49
	(0.24)	(0.26)	(0.22)	(0.25)	(0.51)	(0.50)
Relative capabilities	0.90	0.69	−0.089	−0.52	2.43	1.59
	(1.02)	(1.12)	(0.91)	(1.06)	(2.19)	(2.52)
# of war participants	0.036	−0.049	−0.11	−0.039	−0.72**	−0.52†
	(0.19)	(0.24)	(0.19)	(0.18)	(0.18)	(0.28)
Military strategy	−0.40**	−0.48**	−0.16	−0.21		
	(0.11)	(0.17)	(0.12)	(0.15)		
Major power war	−0.59	−0.96*	−1.50**	−1.69**		
	(0.38)	(0.46)	(0.43)	(0.47)		
Cultural difference	−0.46†	−0.29	−0.57**	−0.93**		
	(0.26)	(0.27)	(0.22)	(0.25)		
Total population			−0.56	−2.12**		
			(0.55)	(0.64)		
Total GDP of Allies					0.41	0.31
					(0.32)	(0.39)
Own GDP					0.36	0.33
					(0.29)	(0.33)
Observations	86	77	86	77	49	49

Note: Standard errors: robust (models 1–4), clustered by war (models 5–6). † p < .1, * p < .05, ** p < .01. Significance tests are two-tailed.

with an increase in the probability of war termination at any point, and hence with a *decrease* in expected war duration.

The single most important finding in table 2.2 is that for the capability shift variable. Larger prewar shifts in relative capabilities, which should be associated with greater worries about future decline and hence more intense commitment problems, are consistently associated with more destructive wars. Thus, in model 1, while controlling for a range of variables that existing literature indicates affect war duration, larger shifts in the years prior to conflict are associated with greater difficulty ending conflicts and hence longer wars. This prediction also holds once I introduce variables intended to capture the effects of other mechanisms of interest in this study, in model 2. Models 3 and 4 shift the focus to battle deaths. Given that a war between Germany and Russia, for example, has the potential to be far more deadly than a war between Guatemala and El Salvador, I include a control for the total population of the war participants. Prewar shifts in capabilities are again associated with a robust and statistically significant increase in total destructiveness. Finally, models 5 and 6 examine total government spending as a share of GDP. Given the limited N, I omit controls for military strategy, major power war, and cultural difference, which are all consistently insignificant once I control for the number of war participants, and I also omit the variable capturing the loser's regime, which in the limited sample of cases for which I have data consists entirely of nondemocratic states (with a maximum Polity score of 1). These regressions do however include controls for the country's GDP and the total GDP of its military allies, both of which plausibly could influence the country's level of spending. Once again, larger prewar capability shifts are associated with greater spending, and hence with an increased deadweight economic loss from war. By any of three different measures of war destructiveness, therefore, when relative capabilities shift dramatically prior to war—a situation that should induce commitment problem concerns—wars tend to be markedly more destructive.

Table 2.2 contains only six of the hundreds of different ways in which we might run each analysis, taking into account possibilities for changing the operationalization of key concepts, the list of control variables, the statistical specification, and the universe of cases. An important question, therefore, is whether these results are representative. While both page limits and the limits to a reader's patience prevent me from presenting results from every possible regression here, I do make results from a wide range of robustness checks available in the supplemental appendix available online, and I summarize them here.

In brief, the results in table 2.2 are strikingly robust. In a minimalist regression with no controls, the capability shift variable misses conventional significance levels, although it is always associated with an increase in war destructiveness. With the standard set of controls, however, the size of the prewar capability shift is *always* significant and in the predicted direction, adopting a range of operationalizations (i.e., capability shifts over five or ten years, for all participants, the primary dyad, or initial participant), whether a different (parametric log-normal) statistical specification is used, whether the universe of cases in several different ways to deal with marginal cases is altered, and whether the Correlates of War data is substituted for the duration and deaths data used here.[44]

In addition to being statistically robust, the relationship between prewar capability shifts and war destructiveness is also substantively quite significant. Building on models 2 and 4, figure 2.1 graphs the survivor function (the probability that war would still be ongoing) for scenarios in which the power shift variable is at either its tenth or its ninetieth percentile, while all control variables are held at their median values. Wars preceded by limited shifts tend to be relatively short and not particularly deadly. Thus, for example, median predicted war duration in this case is under two months, and only about 6 percent of wars are predicted to

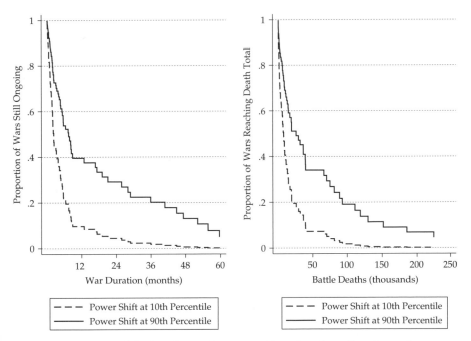

FIGURE 2.1 Fear of decline (commitment problems) and predicted war duration and battle deaths

continue beyond one year. By contrast, in wars preceded by large capability shifts, median war duration is about half a year, more than a third of conflicts last over a year, and roughly a quarter are predicted to last more than two years. Similarly, in wars preceded by limited shifts in capabilities, deaths most often number in the thousands—in only a quarter are battle deaths more than 10,000, and well under 1 percent kill more than 100,000 soldiers—while wars preceded by significant capability shifts are far deadlier, with the median war killing roughly 15,000 soldiers, while about 10 percent kill more than 100,000. These results are quite consistent with the prediction that commitment problem wars will be unusually deadly.[45]

The results in table 2.2 shed some light on the informational and principal-agent mechanisms, as well. When the initiator or the losing side in a war is more democratic, the war tends to be more limited, although this relationship is less robust than it is for the capability shift variable. Specifically, when the losing side in a war is more democratic, the war tends to be both shorter and less deadly—this result is consistent across a wide range of robustness checks. (Recall from above that this variable is omitted from the analysis of spending because of the absence of any remotely democratic war losers in the sample of wars from which there is spending data; if included, the variable is statistically insignificant.) More democratic war initiators, who likewise are conjectured to be constrained from launching particularly open-ended wars, are consistently associated with less destructive wars, however typically at a statistically insignificant level. This relationship is consistent with the argument that greater constraints on democratic leaders mean that they will likely face greater pressure to settle wars more quickly.

The results discussed so far do not provide an adequate test of Goemans's argument that leaders of partially democratic regimes are particularly resistant to settlement, however.[46] Such a test requires the inclusion of a quadratic term to allow for nonlinearities in the effect of regime type on willingness to settle. Table 2.3 thus presents results when including a quadratic term, which is equal to the square of the loser's regime score, alongside the variables in models 2 and 4 of table 2.2. If partially democratic losers are particularly resistant to settlement, we would find that the sign on the linear regime term would become negative, while the quadratic term was positive. Indeed, this is what we observe for the duration analysis (model 1), although the variables are individually insignificant. A graph of predicted effects in figure 2.2 demonstrates that in reality what is occurring is that partially democratic regimes are effectively identical to fully autocratic regimes, rather than that they are unusually prone to extending losing wars. Moreover, in model 2, the results indicate that an increase in the loser's democracy level is

[72]

TABLE 2.3 Effects of losing for partial democracies

	Duration	Deaths
Capability Shift	−1.56**	−1.52*
	(0.54)	(0.61)
Log (war intensity)	0.25*	−0.45**
	(0.11)	(0.13)
Democratic initiator	0.75*	0.58
	(0.31)	(0.38)
Loser regime type	−0.043	0.00086
	(0.069)	(0.072)
Loser regime type^2	0.0049	0.0027
	(0.0035)	(0.0036)
Terrain	−2.39**	−1.4*
	(0.61)	(0.73)
Contiguity	−0.065	−0.14
	(0.25)	(0.25)
Relative capabilities	0.48	−0.62
	(1.07)	(1.06)
# of war participants	−0.023	−0.036
	(0.23)	(0.18)
Military strategy	−0.42*	−0.18
	(0.18)	(0.15)
Major power war	−1.03*	−1.72**
	(0.47)	(0.49)
Cultural difference	−0.27	−0.92**
	(0.27)	(0.25)
Total population		−2.11**
		(0.63)
Observations	**77**	77

Note: Robust standard errors reported. * $p < .05$, ** $p < .01$. Significance tests are two-tailed.

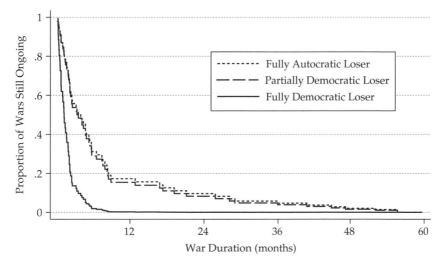

FIGURE 2.2 Losing regime type and predicted war duration

consistently associated with a reduction in total battle deaths at any level of regime type. While signs sometimes change across robustness checks, the basic finding that losing partial democracies are not associated with unusually destructive wars is quite robust. These results thus suggest that there is nothing particularly special about losing partially democratic regimes.

The effect of war intensity varies across specifications, albeit in easily understandable ways. Wars characterized by higher intensity fighting tend to be shorter, as we would expect if the learning permitted by more intense fighting facilitated quicker settlement, but also are unsurprisingly deadlier. The variable is also associated with an increase in total spending, albeit not at a statistically significant level. This variable is of interest primarily for its effect on war duration, which is as predicted. That said, I defer more in-depth discussion of this result to the next section, where I can address potential concerns about whether this relationship actually reflects quicker settlement in more intense wars, as opposed to the quicker military collapse of one side in the war.

The control variables all either perform as expected or are consistently insignificant. Wars fought in rough terrain are consistently longer and, more surprisingly, deadlier. Similarly, military strategy has a strong effect on war duration, with blitzkrieg wars shorter and guerrilla wars longer than conventionally fought conflicts. These strategies, however, have contrasting effects on war duration and war intensity—guerrilla wars, for example, are longer in part because they are fought less intensely— and it is thus unsurprising that military strategy is not a statistically significant predictor of total deaths in war. Wars involving major powers on both sides are also consistently longer and deadlier than other conflicts.[47] Cultural difference is typically associated with both increased war duration and deadliness.[48] Wars involving more populous countries are also typically deadlier, although this relationship is surprisingly statistically insignificant in model 4.

Other control variables are consistently insignificant. Thus there is no clear relationship between contiguity and the duration or deadliness of wars. Similarly, the relative capabilities variable is never statistically significant, and moreover is associated with shorter and less costly but deadlier wars. Finally, an increase in the number of states involved in a war is associated with an unexpected reduction in war duration and at times in deaths, and the variable again is never statistically significant.

Disaggregating Types of War Termination

Many of the hypotheses derived from the different theoretical mechanisms make predictions specifically about the speed with which actors settle their conflicts, rather than the speed with which wars end, which can happen through both settlement and military conquest. Most notably, hypotheses 1c and 1d predict that commitment problem wars will be particularly difficult to settle, but that powers that fear decline face incentives to adopt relatively risky military strategies that produce an increased probability of quick conquest. If these predictions are borne out, we should be more confident that the general relationship between prewar capability shifts and war duration exists for the reasons given by theory. Similarly, the argument that the informational and principal-agent mechanisms are inherently limited focuses on the factors impelling leaders toward settlement, rather than anything about the time at which continued military resistance becomes impossible. In each case, however, general findings could be interpreted as arising primarily because of the relationship between the variable and the speed of conquest. Thus, for example, it is possible that more intense wars would end more quickly because increased levels of fighting encourage faster updating of expectations and hence quicker settlement, but also that intense wars will result in the quick military collapse of one side in the war. Similarly, given the knowledge that a substantial number of democratic losses in war involved European countries conquered by Germany in World War II, it is possible that the finding that more democratic war losers fight shorter wars may not reflect anything about the propensity of democracies to *settle* more quickly.

Adjudicating the effects of variables on the speed of conquest and of settlement thus requires a different statistical specification. This section thus presents competing-risks analyses of war destructiveness. A competing risks specification allows for the possibility that war may end in one of several ways, and thus does not force us to assume, for example, that the end of fighting in Europe in May 1945 reflected Hitler's new-found willingness to reach a mutually acceptable settlement with the Allies. The specific approach adopted here is that of Fine and Gray.[49] The results appear in table 2.4. Starting with the same set of variables as in models 2, 4, and 6 of table 2.2, each regression presents the relationship between variables of interest and the speed of either settlement (models 1, 2, 4, and 6) or conquest (models 3, 5, and 7).[50]

As before, I start with tests of predictions from the commitment problem mechanism. For all three dependent variables, an increase in the capability shift variable is associated with an increase in time until settlement—consistent with the general finding that commitment problem wars tend to be more destructive—but also with a *decrease* in time until conquest. This result indicates that the regressions in table 2.2 were pooling multiple competing effects: on the one hand, declining powers' reluctance to settle meant that ending these wars diplomatically is difficult, but the true size of this effect is undercut by the occurrence of quick conquest in some cases.[51] These wars tend to be more destructive on average (as was seen in table 2.2) because the capability shift variable has a stronger substantive effect on speed of settlement than on speed of conquest and, more important, because most wars end through settlement rather than conquest. Substantively, when holding control variables at their medians, a war preceded by a small capability shift (variable at the tenth percentile) typically reaches settlement in about three months. By contrast, these results suggest that less than a quarter of wars preceded by a significant capability shift (variable at the ninetieth) percentile would end through settlement in the first two years of war. Similarly, in a war preceded by a limited capability shift, the median number of battle deaths prior to settlement is 7,500, a substantial total but no comparison to wars preceded by significant capability shifts, less than a quarter of which are predicted to settle even as battle deaths pass 100,000. Substantively strong and statistically significant results across a number of different dependent variables provide strong support for hypotheses derived from the logic of the commitment problem mechanism.

The results also shed light on the potential alternate interpretations of the war intensity and political regime variables. On the one hand, the finding that regime variables are primarily related to the speed of settlement rather than the speed of conquest demonstrates that the relationship we observe is not simply a consequence of Germany's quick conquest of democratic opponents in World War II but instead is primarily driven by quicker settlement in wars involving more democratic losers.[52] This observation should increase our confidence that the observed effects of regime variables exist because democratic institutions constrain leaders from continuing wars that society would prefer to see ended on terms acceptable to the opponent.

On the other hand, once we focus specifically on duration until settlement, war intensity is no longer a statistically significant predictor of quicker war termination (although the effect is still in the predicted direction). This result is apparently inconsistent with the prediction, drawn from the discussion of the informational mechanism, that high-intensity fighting will encourage faster updating of beliefs and hence produce quicker settlement. If this relationship is not supported, then there is reason to doubt (at least on the basis of statistical findings) the argument that noncommitment problem conflicts are logically limited.

The analysis should not stop here, however. The equifinality of war poses a particular challenge in testing this hypothesis: not all wars are driven by the informational mechanism, and in particular the presence of commitment problem conflicts, which are frequently both intense and difficult to resolve, introduces cases with the opposite relationship

TABLE 2.4 Competing risks regressions of war destructiveness prior to settlement/conquest

	War duration			Battle deaths		Government spending	
	Settle	Settle	Conquest	Settle	Conquest	Settle	Conquest
Capability shift	−3.29** (0.91)	−10.4* (4.52)	3.67** (1.17)	−3.56** (0.95)	3.50** (1.15)	−5.41* (2.61)	4.28† (2.29)
Log (war intensity)	0.11 (0.12)	0.35† (0.18)	0.057 (0.22)	−0.24 (0.13)	−0.10 (0.28)	−0.81** (0.24)	1.81 (1.17)
Intensity × shift		−1.11† (0.66)					
Democratic initiator	0.45 (0.37)	0.41 (0.34)	−0.049 (0.74)	0.070 (0.44)	−0.20 (0.69)	0.82† (0.48)	0.57 (0.95)
Loser regime type	0.065* (0.027)	0.074** (0.027)	−0.015 (0.034)	0.10** (0.036)	−0.033 (0.037)		
Terrain	−1.43* (0.67)	−1.71* (0.68)	1.23 (0.89)	−2.60* (1.11)	1.28 (0.92)	−0.97 (1.46)	1.37 (1.13)
Contiguity	0.15 (0.30)	0.20 (0.31)	−0.091 (0.61)	0.34 (0.31)	−0.14 (0.68)	0.18 (0.60)	−1.61 (1.39)
Relative capabilities	−1.25 (1.27)	−1.58 (1.25)	1.56 (2.28)	−0.58 (1.40)	0.48 (2.22)	−2.81 (2.67)	3.02 (3.52)
# of war participants	−0.22 (0.22)	−0.31 (0.23)	0.46 (0.58)	−0.11 (0.24)	0.55 (0.42)	−0.15 (0.24)	−0.42 (0.38)
Major power war	−0.72 (0.61)	−0.90 (0.66)	−0.41 (0.98)	−0.58 (0.56)	−0.67 (0.96)		
Military strategy	0.14 (0.16)	0.061 (0.18)	−0.71** (0.27)	0.38* (0.17)	−0.51* (0.22)		
Cultural difference	0.56 (0.36)	0.55 (0.36)	−1.58** (0.57)	0.74 (0.47)	−1.56** (0.57)		
Total population				−2.51* (1.23)	−0.27 (1.20)		
Total GDP of Allies						0.65 (0.44)	0.033 (0.32)
Own GDP						0.25 (0.50)	0.12 (0.45)
Observations (Failures due to risk)	77 52	77 52	77 25	77 52	77 25	49 31	49 18

Note: Standard errors: robust (models 1–5), clustered by war (models 6–7). † $p < .1$, * $p < .05$, ** $p < .01$. Significance tests are two-tailed.

between intensity and speed of settlement from that predicted for the informational and principal-agent mechanisms. Figure 2.3 illustrates the problem, focusing for clarity on the commitment problem and informational mechanisms—the expected relationship for principal-agent conflicts would be the same as for informational ones. Assuming a situation in which a third of wars are driven by the commitment problem and are relatively intense, a third are informational and involve intense fighting, and a third are informational but involve less intense fighting, this figure graphically represents the share of each group that theory predicts would still be ongoing at any point in time. High-intensity informational wars end quickly, whereas low-intensity informational and especially commitment problem wars drag on longer. At time t_1, most of the wars that are ending are high-intensity informational conflicts, and thus we would expect a statistical analysis to indicate that increased intensity would be associated with quicker settlement, although the effect would be weakened by the presence of high-intensity commitment problem wars that are not ending. By time t_2, however, all the high-intensity informational wars have ended, leaving only the low-intensity informational wars and the particularly intractable commitment problem wars. With only commitment problem wars represented among the high-intensity conflicts at this stage, we would expect a statistical analysis to reveal that increased intensity is associated if anything with *slower* settlement. Statistical analysis that does not take the equifinality of war into account thus risks concluding that there is no relationship between war intensity and the speed of settlement, or even that the relationship is opposite that predicted by the informational mechanism.

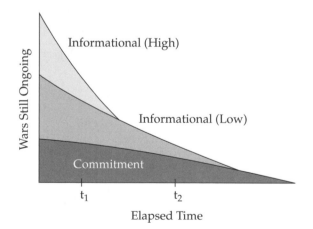

FIGURE 2.3 War intensity and speed of settlement under different mechanisms

[78]

This situation thus implies the existence of an interactive effect: increased war intensity should be associated with quicker settlement when the commitment problem mechanism is not in play, but not when commitment problem concerns are present.[53] The second model in table 2.4 presents results from tests of this prediction. If the interactive argument is correct, then we would expect an interaction term between war intensity and the size of the prewar capability shift to be negative (implying increased time until settlement), while the uninteracted war intensity term would have a positive coefficient. This prediction is indeed precisely what we observe.

For competing risks regression, the most intuitive statistically appropriate method of presenting predicted duration is through the cumulative incidence function, which graphs the predicted proportion of conflicts that have ended in the particular method of interest (here settlement) at any point in time. Figure 2.4 graphs this function for three different scenarios: intense wars with significant commitment problem concerns and both intense and nonintense wars without significant commitment problem concerns. Specifically, the graph presents cumulative incidence functions when prewar capability shifts and war intensity are either one standard deviation above (high) or below (low) their means, holding all control variables at their medians. The central observation here is that the effect of war intensity is highly conditional on commitment problem concerns. When prewar capability shifts are low,

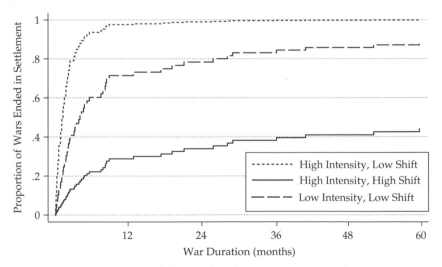

FIGURE 2.4 Cumulative probability of settlement over time under varying conditions

and commitment problem concerns are hence presumably limited, more intense wars reach settlement quickly relative to less intense wars. When commitment problem concerns are salient, however, high-intensity conflicts take a particularly long time to settle. By failing to account for the different relationship between war intensity and the speed of settlement under these cases, the noninteractive analysis in the first model of table 2.4 thus underestimated the strength of support for the prediction that increased war intensity is associated with quicker settlement for wars not driven by the commitment problem mechanism.

Stepping back, this finding is consistent with the most central argument in this book, namely that commitment problem wars are not necessarily limited in the same way that wars driven by other mechanisms are. Wars can be limited either because they are short or because they are not intense. For noncommitment problem wars, we are observing the existence of an intensity-duration trade-off: these results suggest that these wars may be either long or intense, but not both. This trade-off does not, however, apply to commitment problem wars, which thus can account for unusually destructive conflicts.

Finally, the control variables operate in a manner generally consistent with what was observed in the pooled analysis in table 2.2. Rougher terrain particularly impedes settlement, resulting in both longer and deadlier wars. By contrast, shifts away from maneuver toward punishment in military strategy are associated with an unsurprising increase in war duration and battle deaths prior to conquest; this effect is slightly offset, however, by a countervailing reduction in battle deaths prior to settlement. Leaders are also willing to tolerate more deaths prior to settlement when the populations of the war participants are higher. Other variables are either consistently insignificant (contiguity, relative capabilities, GDP variables) or have weak and contrasting effects (number of war participants). Two results are somewhat surprising, however. First, major power wars, although still consistently more destructive, are no longer statistically significant. Second, the competing risks approach reveals that the relationship between cultural difference and increased war destructiveness identified in table 2.2 exists not because difference impedes successful negotiation, as we would expect, but because the amount of destruction prior to conquest is unusually high.

Tests of Ancillary Hypotheses

I am primarily interested in explaining what makes some wars more destructive than others. That said, the discussion of the logic of commitment problem wars identified several predictions that concern other dependent variables. In particular, hypotheses 1b and 1c predict that countries fighting preventive wars driven by shifting power will be more likely to adopt risky military strategies and that these wars will be more likely to end through conquest instead of settlement. Neither of these predictions directly concerns the amount of suffering that a war causes, although the choice of military strategy and the nature of war termination obviously have important political implications. Tests of these hypotheses do however provide a further check on the broader preventive war mechanism: because they follow from the same theoretical argument that predicts that fear of decline will be associated with particularly destructive wars, support for their predictions provides further reason to believe that central argument.

TABLE 2.5 Tests of predictions about war strategy and the nature of war termination, probit regression

	Strategy			Conquest	
	Model 1	Model 2	Model 3	Model 4	Model 5
Capability shift	1.94*	1.99*	2.97**	.39**	2.94**
	(0.96)	(0.99)	(1.00)	(0.75)	(0.92)
Log (war intensity)					0.097
					(0.11)
Loser regime type					−0.038
					(0.029)
Democratic initiator					−0.67
					(0.49)
Democratic involvement		0.32	1.35**		
		(0.47)	(0.46)		
Terrain	0.52	0.65	0.73	0.66	1.21
	(0.81)	(0.86)	(1.00)	(0.70)	(0.76)
Contiguity	0.61	0.67	0.98†	−0.13	−0.18
	(0.47)	(0.46)	(0.59)	(0.37)	(0.44)
Relative capabilities	4.45*	4.51*	1.95	0.43	0.33
	(2.01)	(1.97)	(1.70)	(1.38)	(1.49)
# of war participants	0.11	0.056	−0.033	0.30	0.43†
	(0.23)	(0.23)	(0.21)	(0.21)	(0.24)
Major power war	1.54*	1.72*	1.22†	0.20	0.022
	(0.71)	(0.68)	(0.63)	(0.50)	(0.57)
Prior experience	1.11*	0.99*	0.14		
	(0.44)	(0.48)	(0.48)		
Iron/steel production	−4.60	−6.25	−5.64		
	(5.66)	(5.80)	(7.31)		
Military strategy				−0.46**	−0.42*
				(0.13)	(0.20)
Constant	−6.53**	−6.72**	−5.79**	−0.91	−0.82
	(1.98)	(1.90)	(1.91)	(1.47)	(1.84)
Observations	86	86	86	86	77

Note: Robust standard errors reported; two-tailed significance tests. †p < .1, * p < .05, ** p < .01.

Table 2.5 thus presents tests of both hypotheses. I begin with the prediction that preventive wars will feature unusually risky strategies. While we lack a perfect measure of the riskiness of military strategies, it is generally agreed that maneuver strategies involve a higher than usual degree of risk.[54] By attempting to strike into the rear of the opposing army, attackers using a strategy of maneuver create the possibility of a decisive victory that will free them from having to directly defeat the bulk of the opposing forces. This strategy will be particularly attractive to an attacker who has high war aims and yet may not be particularly overoptimistic about the likelihood of military victory, as I expect will often be the case in preventive wars. The risk, however, is that these strikes typically rely on deception and on the inability of the enemy to respond quickly to the initial attack; an enemy that is undeceived or that redeploys forces quickly may be able to cut off and encircle the attacking force in hostile territory, potentially depriving the attacker of its strongest units. By contrast, both a conventional attrition strategy and a strategy based on punishment, while certainly not free of risk, are much less likely to produce either catastrophic victory or defeat.

I thus test hypothesis 1b by examining when actors choose to use maneuver strategies. The primary variable used in this analysis codes whether either side in a war uses a strategy of maneuver; model 3 focuses by contrast only on cases in which the initiator resorts to maneuver warfare. In addition to the standard controls used elsewhere in this chapter, I also include controls for several variables—the level of economic development, democratic involvement in the war, and past experience with maneuver warfare—that were found to be significant predictors of maneuver strategies in a previous study by Reiter and Meek.[55] The proxy for economic development uses the level of iron/steel production from the National Military Capabilities dataset.[56] The measure of past experience with maneuver is intended to capture a hypothesis based on learning. The variable is coded 0 if the primary participant on a side has had no direct experience with maneuver warfare over the previous twenty-five years. It is coded 1 if the country successfully used maneuver warfare or had maneuver warfare used successfully against them, and -1 if it used maneuver unsuccessfully or if it defeated an opponent who used a maneuver strategy.[57] The democratic involvement variable measures the presence of a democracy on either side, to capture the argument that democracies are more likely to use maneuver strategies. In all cases, the variable used combines data for each side in the war (by summing the variables for economic development and past experience, and by coding democratic involvement as 1 as long as at least one democracy was involved); substituting the attacker's data alone produces effectively identical results, however.

Model 1 of table 2.5 presents a probit regression of military strategy on the prewar shift in capabilities and a set of standard controls. Consistent with the argument that preventive motivations for war tend to be associated with the use of riskier strategies, prewar shifts in capabilities are associated with a significant increase in the probability that one side in the war uses a maneuver strategy. Model 2 adds a control for democratic involvement in the war. The prewar power shift variable is again statistically significant. In both these regressions, most controls behave as expected, although not always at a statistically significant level.[58] Model 3 restricts the dependent variable to cases in which only the war initiator uses a maneuver strategy. Again the power shift variable is statistically significant. Using model 2 and holding control variables at means or medians, the predicted probability that a maneuver strategy is used rises from about 5 percent when the prewar power shift is at its tenth percentile to about 35 percent when it is at its ninetieth percentile, roughly a sevenfold increase in likelihood.

Models 4 and 5 test the prediction that preventive wars driven by fear of future decline will be more likely to end in conquest, as opposed to settlement. Model 4 presents the results with baseline controls; model 5 adds in war intensity and regime variables. Larger shifts prior to war turn out to be strongly associated with an increased probability that war ends through conquest. This result holds for a range of alternate specifications, including substituting the more expansive and more restrictive codings of conquest described above. Using model 5 and holding other variables at means or medians, an increase in the power shift variable from its tenth percentile to the ninetieth percentile is associated with an increase in the predicted probability of conquest from just over 5 percent to more than 50 percent. These results are strongly supportive of the prediction that preventive motivations driven by shifting power will be associated with an increased probability that wars ends through conquest instead of settlement.[59]

The statistical analysis in this chapter is quite supportive of the central claims of this book. Most significantly, I find that high prewar shifts in relative capabilities, which I argue proxy for anticipated future shifts (either through a continuation of current trends, as with the US-China relationship today, or through a reversion to the mean that closes an existing window of opportunity, as with the German-Soviet relationship after the purge of the Red Army), are associated with a consistent increase in the destructiveness of war. This result holds for three different measures of war destructiveness—duration, battle deaths, and government expenditures—and is robust to a wide range of changes to the analysis, including using different operationalizations of prewar power shifts, altering the universe of cases, or changing the statistical specification. These findings are strongly consistent with the prediction that fear of decline is associated with unusually destructive wars.

Moreover, further analysis demonstrates support for additional hypotheses that follow from the same theoretical arguments that predict that commitment problem wars will be unusually severe. In particular, the commitment problem mechanism creates incentives for leaders who fear decline to resist settlement but also to adopt risky military strategies that increase the likelihood that either they or their opponent will experience military collapse. Consistent with this argument, competing risks regressions demonstrate that prewar capability shifts are associated with a dramatic increase in the time that it takes for settlement to arise in war, but also with a significant reduction in the time that it takes for wars to end through conquest. Probit analysis of whether war participants adopt maneuver strategies, which generally are riskier, and whether the war ultimately ends through settlement or conquest, produce results that are similarly consistent with theory. The statistical analysis thus suggests not just that commitment problem wars are unusually destructive, but that they are destructive for the reasons that theory leads us to expect.

The statistical analysis also provides support for theoretical arguments about limited wars. Most important, consistent with the arguments about limited wars, when fighting is more intense, and both leaders and the people they represent thus have greater opportunities to update their beliefs, settlement occurs more quickly. Demonstrating this relationship required separating out (by inserting an interaction between prewar shifts and war intensity) the countervailing effect of commitment problem wars, which tend to be both intense and difficult to settle. This finding suggests that noncommitment problem wars can be either intense or long, but not both, implying in turn that the informational and principal-agent mechanisms provide reasonable accounts for limited wars but not for unlimited ones.

Similarly, wars involving more democratic initiators and losers tend to be more limited, consistent with the argument that increased institutional constraints on leaders limit their ability to continue wars unnecessarily. At the same time, I find no support for prior arguments, which focus on leaders' incentives rather than on constraints, that losing partial democracies are particularly resistant to settlement. Although these findings constitute less direct tests of my arguments than others presented here, they are consistent with the claim that leaders face significant constraints on their ability to continue wars that are not in the interest of a substantial portion of society, which in turn provides further reason to expect that principal-agent wars will be limited.

Overall, then, these results provide strong support for the central arguments in this book. That said, the findings in this chapter are certainly not beyond all doubt. Thus, for example, prewar shifts in capabilities are an admittedly imperfect proxy for fears of future decline. Moreover, all these analyses require me to test hypotheses drawn from specific mechanisms on cases in which those mechanisms likely were not active. In this context, then, these results will gain strength to the extent that they are bolstered by evidence from specific cases from history. The next five chapters thus shift to a series of case studies, which allow me to test hypotheses more directly, if on a smaller sample of cases. These case studies will also permit me to test claims about dispositional commitment problems and unconditional surrender, which could not be tested quantitatively given the limited number of cases in which this behavior has arisen throughout history. Thus, while the statistical analysis in this chapter provides reason to believe that the theoretical arguments in this book are correct, the evidence is as yet incomplete—the remainder of the book aims to fill in the remaining gaps.

[3]

War to the Death in Paraguay

The war between Paraguay and its neighbors, which was fought from 1864 to 1870 and which ended with the utter destruction of Paraguay, is one of the great unknown wars in history, little studied despite being at a per capita level quite possibly the deadliest conflict in the past two centuries.[1] While disagreement exists about the exact numbers of dead, it is undeniable that Paraguayan human losses were huge, by some estimates as high as two-thirds of the prewar populavtion.[2] For outsiders, the sketch of the war makes the Paraguayan president—Francisco Solano López—appear frankly insane. As leader of a small buffer state whose independence was far from assured or even fully acknowledged, he attacked both Brazil and Argentina, the great powers of South America. He then refused to back down, fighting against worsening odds for six years, in the process dragging his country to ruin. Traditional histories generally portrayed López as a psychopath, a nineteenth-century Hitler determined whatever the odds to turn his country into an empire or to destroy it in trying.[3] Such a man seems well outside the ken of rationalist theories of war.

Closer examination, however, shows that López had clear and reasonable justifications for engaging in the war that he undertook. While the course he chose was risky, he believed for understandable reasons that inaction was also quite risky. Brazil posed a constant creeping threat to Paraguay's survival, while Argentine leaders in Buenos Aires had never truly reconciled themselves to Paraguayan independence. Historically, Paraguay had survived, like most buffer states, by playing off the two powers against each other, but the rapprochement of Argentina and Brazil, associated most strongly with Bartolomé Mitre's acquisition of power in Argentina, fundamentally reshaped power politics in the region.

Brazil's intervention in Uruguay, supported by Argentina and justified on flimsy pretexts, was not unsurprisingly interpreted as the first step in the annexation of Uruguay and Paraguay. Paraguay's only real access to the outside world—essential for supplies both military and mundane—was through the various rivers that fed into La Plata; all ran through Argentina, Uruguay, or both. The loss of a Uruguayan ally thus weighed more heavily than Uruguay's slight capabilities might have indicated.

Had he recognized how the war would go, López likely would not have chosen to start it. However, the decision to attack was less insane than it at first appears. Paraguay by most accounts had the best army in the region and thus had a reasonable chance of defeating the Brazilians in open battle. Moreover, one of the apparently most inexplicable decisions of the war—attacking Argentina when already at war with Brazil—becomes much more understandable when we recognize that it was far from certain that all Argentina would rally against him. The country was deeply divided, and influential figures—most notably Justo José de Urquiza, the governor of Entre Ríos Province—had more in common with López than they did with Bartolomé Mitre in Buenos Aires. An alliance with important Argentine Federalists was definitely a possibility, and had it occurred the war could easily have followed a very different path. When he launched his attack, therefore, López was less crazy than in retrospect he appears to have been.

Given the strategic problem posed by the gradual increase in Brazilian and Argentine ability to project power into the interior and the closer relations between the two traditional rivals, López needed a significant victory to force his opponents to abandon their claims to Uruguay and to force a final delineation of the disputed border. When his initial attack miscarried, therefore, he was willing to continue to fight. Once the Allies forced their way onto Paraguayan territory, however, it was clear that the best possible military result for Paraguay would be a successful defensive war, which would not address the sources of Paraguay's strategic decline. It was only at this point, then, that he began to demonstrate willingness to negotiate on terms in line with the military situation on the ground. This behavior accords closely with what we would expect were a fear of future decline to be the primary motivation for war.

Yet here the logic of the dispositional commitment problem intervened to assure a war to the finish. Deeply offended by Paraguayan policy, key Brazilian officials, central among them Emperor Pedro II, had concluded that a viable peace required the expurgation of López and the entire system of *caudillo* politics that he represented. Indeed, this case provides crucial evidence for the unconditional surrender mechanism. The Brazilians, who appear not to have intended to do what López feared, concluded that the Paraguayan leader was a menace with whom negotiation was impossible, whereas the Argentines, who actually did want to annex Paraguay, understood his motivations and hence were willing to negotiate. In the end, convinced of López's iniquity, the Brazilians rejected entreaties from all quarters to negotiate, instead carrying the war forward at high cost until the Paraguayan defenses crumbled, Asunción was occupied, and López himself fell on the field of battle.

The History and Anomalies of Paraguayan Policy

Under the Spanish Empire, Paraguay was part of the same viceroyalty as what ultimately became Argentina, with effective power centralized in Buenos Aires. Following independence from Spain during the Napoleonic period, the Paraguayans resisted attempts to bring them under the auspices of a new government in Buenos Aires and declared independence, although no power recognized them as independent for thirty years.[4] Indeed, at this time and continuing through the Paraguayan War, it was unclear which regions (if any) would remain independent and which would be absorbed under the control of Buenos Aires or Brazil. Paraguay and Uruguay ultimately emerged as independent (if highly constrained) states from this process, but others in similar positions—such as Corrientes, Entre Ríos, and Rio Grande do Sul—did not.[5] The new Paraguayan state had disputes, including large border disagreements, with both its larger neighbors, and following the classic strategy of a buffer state, it played each against the other to secure its own independence.

This approach worked well when both Buenos Aires and Rio de Janeiro lacked the capacity to project significant power into the South American interior, and it remained functional even when that guarantee vanished so long as the Paraguayans could ensure that each of their larger neighbors would oppose any unilateral move by its rival. Paraguayan strategy was undermined, however, by a rapprochement between Brazil and Argentina that began with Bartolomé Mitre's accession to power in Buenos Aires in 1860. The primary manifestation of this rapprochement was agreement on affairs in Uruguay, where liberals (*Colorados*) under Venancio Flores had rebelled in an effort to overthrow the conservative *Blanco* regime. The Argentines stood by passively, even aiding Flores behind the scenes, while the Brazilians provided increasing support, ultimately intervening directly. The Blancos appealed to López for assistance, and he remonstrated with the Brazilians, but to no effect. At this point, therefore, he decided to go to war.

Geography posed a significant obstacle to direct assistance, however, as figure 3.1 illustrates. Paraguay and Uruguay lacked a common border,

Caracas
VENEZEULA
Bogota
COLOMBIA
GUIANAS
Quito
ECUADOR
PERU
Lima
BRAZIL
Salvador
La Paz
BOLIVIA
PARAGUAY
Rio de Janeiro
São Paulo
Asunción
Corrientes
Santa Fe
Paraná
Córdoba
URUGUAY
Santiago
Buenos Aires
Montevideo
ARGENTINA
CHILE
PATAGONIA
SOUTH AMERICA
circa 1850
N
0
500 miles

FIGURE 3.1 South America prior to the Paraguayan War. Reprinted from *The Paraguayan War, Volume I: Causes and Early Conduct* by Thomas L. Whigham by permission of the University of Nebraska Press. Copyright 2002 by Thomas L. Whigham.

and passage through Brazilian territory would have been extremely difficult.[6] López's first move was therefore to attack northeast into the Brazilian province of Mato Grosso, which allowed him to capture a large cache of arms and had the potential to divert the Brazilians from their interference in Uruguay. His incursions succeeded militarily but failed strategically, as the Brazilians did not significantly oppose the Paraguayan advance into the province, instead pressing the attack on Montevideo in Uruguay. López then turned south and requested permission from Buenos Aires to cross the province of Corrientes, which was frequently rebellious but generally acknowledged to be Argentine territory. Mitre in Buenos Aires refused, prompting López to declare war. By this point, however, the Brazilians had taken Montevideo and installed Flores in power; the Paraguayans advanced down the Uruguay River to reverse the decision, but ultimately failed and were forced back to their territory.

In response to the Paraguayan attack, Brazil, Argentina, and a now-occupied Uruguay formalized their alliance through a treaty signed on May 1, 1865 (hence the common name "the War of the Triple Alliance"). In this secret alliance, Brazil and Argentina each agreed to permit the other to satisfy the full extent of its claims on Paraguayan territory. Achieving these goals required that the Allied forces advance into Paraguay, a task that proved far more difficult than expected, as rough terrain and stiff Paraguayan resistance slowed progress. After a buildup, the Allies forced their way across the Paraná River into Paraguayan territory in April 1866. The subsequent advance toward the main Paraguayan defensive fortifications at Humaitá was slow and costly, and it stopped completely following a significant Allied defeat in September at Curupaity, which, when combined with domestic unrest, forced Argentina to withdraw the bulk of its army from the war effort. The scale of this defeat forced the war into a pause of almost a year during which the Allies brought forward reinforcements; during this time, the front did not move, although Paraguayan forces constantly harassed their opponents. The Allies began to move again in August 1867 and, after hard fighting, eventually circumvented the Paraguayan defenses and captured Humaitá; its fall doomed Asunción, which fell on January 5, 1869. At this point, López transitioned to a strategy of guerrilla war, reforming his army (which by then consisted almost entirely of the very old and the very young) and fighting the Brazilians on the run. The war thus continued for more than a year, ending only when Brazilian forces tracked down and killed López on March 1, 1870.

That the war continued this long militarily was a function of difficult terrain, Allied disorganization and strategic errors, and the conviction and ability of Paraguayan soldiers. Yet politically there were many opportunities to end the war far earlier. Although international involvement was limited, American representatives made a number of serious attempts to mediate, while representatives of European powers made some more half-hearted tries. Moreover, at several points in the later portion of the conflict, including a well-known meeting at Yataity-Corá just before the Battle of Curupaity as well as in response to several mediation offers, López indicated a willingness to negotiate. Throughout the war, however, the Brazilians consistently rejected all talk of negotiations or peace, even on quite advantageous terms. This resolution to carry the fight through to the finish was not shared by the Argentines or the Uruguayans, and indeed after Curupaity the two countries withdrew all but a token force. In the end, therefore, the conflict became a war to the death between Paraguay and Brazil.

Ultimately, this case contains a number of important puzzles. Why would López pick a fight with the region's greatest power when his ability to defeat his opponent was quite questionable? Why, having done so, would he then decide to go to war against the other regional power as well? Why would he continue to fight even after the fall of Uruguay, when the military odds were stacked against him? Why, in other words, would he precipitate and then fight a serious war in which Paraguay's chances seemed so poor? On the other hand, once López began to demonstrate a greater willingness to negotiate, why were the Brazilians so implacably opposed to talks, especially considering the high costs and limited material benefits of a war to the finish?

As I detail below, the answer to the questions about Paraguayan intentions hinged in part on the fact that López had reasons for optimism that are not immediately apparent, largely because some of them did not work out the way that he (not unreasonably) expected. A larger part, however, was structural: Paraguay was in a dangerous position that threatened only to get worse, and given reasonable fears about the intentions of his opponents, López decided that Paraguay's choice was between fighting an unattractive war or waiting for its neighbors to partition the country at a later date. In this context, he was willing to accept even a fairly risky and costly war, given what he saw as the alternative. He failed, however, to anticipate the nature of the Brazilian response, and as a result the war that he got resulted in the utter destruction of Paraguay.

DIVERGENT EXPECTATIONS: NECESSARY BUT NOT SUFFICIENT

One of the more obvious puzzles of the Paraguayan War is why López decided to challenge either one of South America's great powers, let alone both. While the decision to start a war against both Brazil and Argentina has led some to question López's sanity, the Paraguayan leader's behavior was less utterly inexplicable than it initially appears. Paraguay was far more capable than would have been expected from its isolated geography and relatively small territorial size. Moreover, López started the war with the ultimately incorrect but hardly unjustified belief that he would have important allies on his side. From this perspective, the informational mechanism has at least a partial claim to explaining the decision to start the war.

Consider first Paraguayan capabilities.[7] Under Carlos Antonio López, the father and predecessor to Francisco Solano López, the country moved away from its previous autarchy, importing both goods (including weapons) and advisers from Europe. When combined with the imposition of a universal draft, the creation of a domestic capacity to produce munitions, and a strong state role in the economy that facilitated extraction, these policies meant that Francisco Solano López inherited a surprisingly capable army upon assuming power.[8] By contrast, Paraguay's two main opponents were unprepared for war. While Brazil had a huge national guard, that force was poorly equipped, untrained, and effectively useless—when the war began, the government created a new army rather than rely on the guard—and the standing army was a small, ill-trained force that consisted largely of press-ganged vagrants.[9] Moreover, the conservative elites who dominated the country viewed a strong army as a threat to the existing social order. In Argentina, there was no national army, as the political divide between Buenos Aires and the provinces (discussed in greater detail below) was paralleled in the military. While Mitre was taking steps to develop a national army centered on Buenos Aires when the war began, this process had only just begun. Uruguayan forces, on either side of the Blanco-Colorado political divide, were experienced but few in number and poorly trained. Overall, therefore, Paraguay had the largest and best-equipped army in the region, while its opponents would need more than a little time to make up the gap. Indeed, in the opening stages of the war, Mitre apparently worried that the invading Paraguayan armies would join up at the junction of the Uruguay and Paraná rivers to directly threaten Buenos Aires.[10]

As for the apparently inexplicable decision to provoke a war with Argentina when already at war with Brazil, here hindsight is a hindrance to our ability to accurately understand López's decision. In brief, he had good reason to believe that were Argentina to get involved in the war much of the country would fight on Paraguay's side. The country was split between Federalists, led by Justo José de Urquiza of Entre Ríos province and disproportionately populated by the rural caudillos, and centralists led by Mitre and concentrated in Buenos Aires, although also present elsewhere. Attempts immediately after independence to impose a centralized government run from the capital encountered significant resistance that ultimately forced the adoption of a federal constitution. For more than twenty years, Juan Manuel de Rosas effectively ruled the country as governor of Buenos Aires, but he had to contend constantly with regional caudillos and ultimately angered enough people domestically and internationally to generate a coalition that rallied behind Urquiza to overthrow him.[11] For six years after the overthrow Buenos Aires was legally independent from the rest of Argentina, until Urquiza, recognizing that the provinces were dependent on revenue from the main city's trade, invaded in 1859 and reestablished the political link. When Mitre took power as governor of Buenos Aires in 1860, therefore, he was in a decidedly secondary position. It was only when he successfully fought off yet another invasion by Urquiza in 1861 and then was elected the first president of the Argentine Republic in 1862 that he had any serious claim to ruling the country, and even then he could not count on Urquiza's loyalty. To say that López attacked Argentina is thus to give the country greater fixity than it actually possessed at the time.

This situation was integral to López's planning. When he requested permission to cross Corrientes, he believed that, should war happen, Urquiza would side with him. Indeed, he had good reason for believing this, as Urquiza had said he would: during the Mato Grosso campaign, the caudillo allegedly told López's representative that "he favored Paraguay in its war against Brazil and that if it were to go through Corrientes to invade the empire, he would personally come and offer himself as a volunteer."[12] Moreover, Urquiza's followers, and even his son, were incensed with Mitre's Uruguay policy and thus were pressing him to back Paraguay.[13] While Urquiza began to backtrack prior to López's request, he could not undo the effect that previous signals of support had had on López's calculations.[14] Even after Urquiza's loyalty to Mitre became clear, internal unrest posed a significant obstacle to Argentina's vigorous prosecution of the war.[15]

Overall, therefore, to use the words of one of the foremost experts on the war, "the marshal's plan was ambitious but not insane."[16] López was clearly gambling, and arguably gambling at fairly long odds, when he took his country to war against the two foremost powers of the region, but it was not wishful thinking for him to have hoped that his gamble might pay off. He was ultimately mistaken in his expectations, particularly with respect to Urquiza, and had he better understood the situation in Argentina he probably would have been more cautious in taking his country to war. It is thus impossible to account for his behavior without recourse to the informational mechanism.[17]

[93]

That said, there are significant limits to the informational mechanism. To say that López's plan was not insane is not to say that it was necessarily a good idea. If López had reason for optimism, a war with Brazil was nonetheless a significant risk, and his declaration of war on Argentina likely made it easier for Urquiza to stay loyal. Moreover, throughout the period of the Paraguayan offensive against Brazil and Argentina, López did not articulate clear political demands of the sort that his opponents might conceivably agree to, nor does he appear to have scaled back his ambitions in response to the unfavorable revelation that Urquiza would not back him. In addition, the decision to push Paraguayan troops far outside his country's territory entailed significant risks that, from the perspective of the informational mechanism, were unnecessary. Simply holding Mato Grosso, the Brazilian province captured in the initial months of the war, might well have been sufficient to induce Brazil to negotiate a final delineation of the two countries' borders, long a Paraguayan objective.[18] In all, based solely on expectations, it is hard to understand why López felt the need to launch a risky military adventure to assist an ally (the Uruguayan Blancos) of dubious loyalty and minimal capacity when a more limited policy might have achieved significant gains, nor can the informational mechanism explain why he did not try harder to escape from the war as Paraguay's diplomatic situation deteriorated. To understand these decisions, we must turn to the preventive war mechanism.

Paraguay's Perilous Position and Fears of the Future

Ultimately, an explanation for López's ambitious foreign policy and his willingness to run acknowledged risks must take into account the difficult, and worsening, position in which Paraguay found itself. The basic problem was that the country was surrounded by expansionist neighbors whose refusal to guarantee Paraguay's borders or at times even acknowledge its independence prompted fears that any passive policies would merely acquiesce in the gradual dismantling of the country. Traditionally, and again after the war, Paraguayan leaders had dealt with this problem by playing the two powers off against each other. Mitre's rise and the subsequent Brazilian intervention in Uruguay fundamentally undermined this strategy and raised the specter of a blockade of the Paraná River—Paraguay's outlet to the sea—that by cutting off access to Europe would slowly strangle López's country and set the stage for an eventual partition. In this context, continuation of current trends threatened Paraguay's survival, while war provided the only available option to upset those trends. This interpretation thus explains why López

[94]

adopted such a risky strategy and made such apparently extreme demands. Ultimately, of course, Paraguay failed militarily and had to revert to a defensive posture. Consistent with the preventive war argument, López refused to negotiate until it was clear that there was absolutely no hope of achieving the kind of victory that would be necessary to prevent Brazil from consolidating its position in Uruguay.

To understand the dangers that López confronted, it is necessary first to briefly review his country's history. In the final decades of the Spanish Empire, Paraguay was united administratively with Argentina and Uruguay under the Viceroyalty of the Río de la Plata. In this context, Buenos Aires used its favorable geographic position to impose shipping taxes that stifled Asunción (as well as the Argentine interior).[19] Following the break with Spain in 1813, the Paraguayans established an effective independence that was, however, only tenuously acknowledged by its neighbors, who advanced significant territorial claims that were the basis for persistent border disputes. Repeated attempts to resolve the disputes under Francisco Solano López's predecessors—José Gaspar de Francia and Carlos Antonio López—ended in failure, with the Paraguayans typically given little evidence that opposing interlocutors were willing to negotiate in good faith.[20]

The Brazilians, who inherited the Portuguese strategy of encouraging settlement to subvert legal border agreements by changing the facts on the ground, were pushing settlers into Mato Grosso, an isolated interior province that was almost impossible to reach except along the river through Paraguay.[21] From there, geography provided a clear motive for a policy of gradual expansion: "from the Brazilian point of view, Paraguay was the natural extension of Mato Grosso, and any geographer could see that the republic stood poised like a dagger at the entrails of the empire."[22] Meanwhile, the Paraguayans were exercised by the shipment of cannons into Mato Grosso that, while in reality of limited military utility, provided apparent evidence of hostile intentions.[23] López could thus be forgiven for concluding that inaction merely invited the Brazilians to subvert his country's independence gradually.

Meanwhile, Paraguay and Argentina disputed ownership over a large, if sparsely populated, region. The real danger here, however, was that many in Argentina, Mitre among them, had never reconciled themselves to Paraguayan independence and hoped to reunify the disparate territories of the old Spanish viceroyalty in a large Argentine state that would constitute a true rival to Brazil for hegemony in South America.[24] In this context, a centralized government dominated by Buenos Aires would pose the greatest threat (influential provinces like Entre Ríos and Corrientes frequently shared interests with Paraguay and indeed contemplated independence).[25] As head of the *porteños*—the inhabitants of

Buenos Aires who tended to favor a strong, centralized government—Mitre thus posed a threat purely from the perspective of Argentine-Paraguayan relations.

That said, the threat posed by a centralized Argentine government was dwarfed by the danger associated with Mitre's improved relations with the Brazilian Empire.[26] Paraguay was a buffer state, and buffer states—generally unable to guarantee their independence through force of arms alone—have always had to rely for their survival on their ability to manipulate tensions between their stronger neighbors. Should such tensions vanish (or should the rivals work out an equitable plan for partition), even simple survival cannot be guaranteed; it is thus unsurprising that buffer state status is the strongest predictor of state death in the international system.[27] Geopolitically a buffer state, Paraguay was however unusual in that in 1864 it had the means plausibly to do something to address the problems created by an Argentine-Brazilian rapprochement.

Internal Argentine divisions, discussed with reference to López's bases for optimism above, were at the center of the issue. Mitre's hold on power was threatened by opposition from the Federalists, many of whom felt greater affinity to the Paraguayans than they did to the citizens of Buenos Aires. Thus, for example, Ricardo López Jordán, who ultimately would assassinate Urquiza in 1870 in large part in response to his policies in the war, appealed to him at the time, saying, "You call us up to fight in Paraguay. Never, General: those people are our brothers. Call us up to fight against *porteños* or Brazilians. We are ready. They are our enemies. . . . This, I'm sure, is the true sentiment of the *entrerriana* people."[28] Given these internal divisions, Mitre saw an alliance with the Brazilians as a means to cement his authority and that of his faction. This political situation thus served as the basis for cooperation in Uruguay, where Brazil intervened directly while Argentina funneled weapons and supplies to Flores. From Paraguay's perspective, of course, a strategy of playing the two rivals off each other ceased to work if the rivals became friends. López had two options available to him: to wait and hope that opposition within Argentina to cooperation with an old enemy would bring about Mitre's fall, or to take action to break up the incipient alliance and perhaps precipitate that fall directly. Both options were risky; he chose the latter.

REASONS AND AIMS: LÓPEZ'S DECISIONS FOR WAR

The final piece of the puzzle was the Brazilian intervention in Uruguay. British intervention in the Cisplatine War of the 1820s, undertaken to protect trade throughout the region, was the original reason why Uruguay attained independence.[29] Uruguay's independence thus benefited Paraguay, as it complicated any attempt to bottle up Paraguayan trade and contact with the outside world. Indeed, over time Paraguay became increasingly dependent on international trade, both for economic development and for the military supplies that would be needed to fend off encroachments into Paraguayan territory.[30] When Argentina and Brazil turned away from their traditional rivalry to jointly back Flores and the Colorados in Uruguay, however, the specter of encirclement came fully into view. With Uruguay under hostile control, there was little that the Paraguayans could do to prevent Argentina and Brazil from simply shutting off commerce to Asunción to slowly but surely strangle Paraguay, especially as the Argentine occupation and fortification of the island of Martín García in the context of the Uruguayan intervention increased the direct control that the Argentines could exercise over the Paraná River delta.[31]

Nor did the participants do much to allay Paraguayan fears. The Brazilians never provided a satisfactory explanation for the buildup of military forces in the Mato Grosso, nor did either Brazil or Argentina give good reason why a final delineation of the borders was impossible.[32] When López sent his ultimatum to the Brazilians, the Brazilian ambassador in Asunción took it on himself to reply snippily that Paraguay's involvement in the Uruguayan affair was not needed.[33] Caricatures of López that appeared in the press in both Buenos Aires and Rio de Janeiro certainly did not help: from his perspective, not only did his neighbors ignore his legitimate concerns; they belittled him for raising complaints.[34] All of these developments could only have added to López's concern that, if Paraguay wished to remain independent, it would need to act quickly.[35]

To the extent that they have survived, López's public and private statements are consistent with the view that a fear of future decline drove his policy decisions. In his ultimatum to Brazil, he closed by stating that

> the government of the Republic of Paraguay will consider any occupation of the Oriental territory [i.e. Uruguay] as an attempt against the equilibrium of the states of the Plate which interests the Republic of Paraguay as a guarantee for its security, peace, and prosperity; and that it protests in the most solemn manner against the act, freeing itself for the future of every responsibility that may arise from the present declaration.[36]

Likewise, one of López's European advisers reported that, when deciding whether to seize the Brazilian ship *Marqués de Olinda*—the first Paraguayan act of war—López averred that "if we don't have a war now with

Brazil, we shall have one at a less convenient time for ourselves."[37] More generally, throughout the war he accused the principal Allies of aggression despite the fact that neither of them had directly attacked him.

Paraguayan war aims are similarly consistent with what we would expect from the preventive motivation for war. Although significant dispute exists as to exactly what López hoped to achieve, his minimal aims seem clear. By invading down the rivers, he hoped ultimately to reach Uruguay, where he would then restore the Blancos to power. This action would restrict the Brazilian and Argentine influence in Uruguay, preventing the anticipated shift in capabilities that their domination over Uruguay would ultimately imply. This goal was audacious: Paraguay did not share a border with Uruguay, and thus the Paraguayan army had to advance through several hundred miles of non-Paraguayan territory and then reverse the status quo by taking a city that not many years earlier had successfully withstood a nine-year siege.[38] This plan thus involved sending his best units far forward, where they were in danger of being cut off and forced to surrender, as indeed ultimately happened. As a military plan, it was thus far riskier than a more defensive strategy, such as seizing Mato Grosso and then waiting to repel the Brazilian counterattack. Only this aggressive strategy held out any prospect of dealing with the problems that motivated López to start the war in the first place, however: if Paraguay simply occupied Mato Grosso and waited, nothing would prevent Brazil from simply waiting until a blockade had crippled Paraguay, or from reaching an agreement only to abrogate it later.

As long as the phase of Paraguayan offensive operations continued, López never raised the prospect of negotiations. The naval defeat at Riachuelo and the subsequent capitulation of the forward armies, most notably the surrender at Uruguaiana in September 1865, blunted the Paraguayan ability to attack, however, and once the allies forced their way across the Paraná into Paraguayan territory López had to admit that his war aims were no longer achievable. It was thus not until 1866 that he began to show an interest in negotiations. Historians have debated the degree to which these offers were sincere, with both detractors and admirers of López frequently arguing that they were intended only to buy time for his army to build up their defenses.[39] Yet there is good reason to believe that he was genuinely interested in peace, should peace be available on reasonable terms. He was the leader who proposed the discussions at Yataity-Corá, and in the discussions he was willing to entertain a range of proposals, although the refusal of the Brazilian representative to talk (and the Allied refusal to scale down territorial demands) constituted an insuperable bar to any agreement.[40] At later points, he was willing to discuss specific bases for peace with American and European representatives who might serve as mediators with the Allies.[41] What he was not willing to do was to sign over his country's independence when the Allies were not yet in a position to compel him to capitulate. Aware that Paraguay was bottled up with no prospect for escape, he nonetheless sought to use Paraguay's still-considerable capacity to resist invasion to save his country at least for the time being. His position at this point was quite reasonable: the Allied territorial conditions went well beyond what they could reasonably demand given the military situation prior to the fall of Humaitá in late 1868. Throughout this period, therefore, Paraguayan policy was consistent with what we would expect from a country responding to an anticipated adverse shift in relative capabilities.

Summary of a Preventive War

The puzzles of the length and destructiveness of the Paraguayan War are in many ways closely connected to the puzzles of its onset. Why did the leader of a buffer state start a war with not just one but two regional powers? Why adopt a risky and aggressive military strategy when a more limited approach would have increased the challenges for his opponents? Why were the Paraguayans and López willing to fight so fanatically, literally to the death? The willingness to adopt an aggressive military strategy and to countenance high costs in the pursuit of victory was closely connected to the extraordinary destructiveness of this war. As with any war, the answer to these questions is nuanced, but the broad theoretical predictions from chapter 1 are borne out quite well. A full understanding of López's strategy must draw on both the informational and the preventive war mechanisms. The informational dynamic is particularly pertinent for López's decision to attack Argentina when already at war with Brazil: given the expectation that Urquiza and much of rural Argentina would rally to his side, it was far from clear that the invasion of Argentina would greatly worsen Paraguay's strategic position.[42] The aggressiveness and urgency of Paraguayan policy can be understood only given the understanding of Paraguay's perilous and, more important, worsening strategic position. Intervention to prevent Brazilian and Argentine domination of Uruguay was reasonable given the inference that Paraguay was a logical next target and that undermining Paraguayan independence would be easier once Brazil and Argentina were in a stronger position cut Paraguay off from the outside world. Moreover, war with Argentina, by provoking civil unrest in that country, had a good chance of unseating Bartolomé Mitre and thus addressing the primary source of Paraguay's strategic decline, which was the rapprochement between Brazil and Argentina under Mitre. In this context, an aggressive war, even at relatively long odds, made sense. It was this strategic setting that provoked López into launching an unusually aggressive and risky war and that helped to ensure that Paraguayan resolve would be so high: although the aggressor in this conflict, the Paraguayans sincerely believed that they were fighting for their independence.

WAR TO THE DEATH

An explanation for why and how the war started and the way in which it was initially waged starts, therefore, from the Paraguayan fear of decline. Yet Paraguayan fears, and the policies that followed from them, cannot account fully for the way in which this war ultimately was fought. By the midpoint of the war, López began to indicate a willingness to settle the war on terms that were reasonable given the military situation. At this point, then, we might have expected to see the war move to a close. Indeed, on behalf of Argentina, Bartolomé Mitre repeatedly indicated his willingness to negotiate, both in talks with López and in discussions with potential mediators.[43]

The Brazilians, however, were resolutely opposed to peace. In November 1865, Emperor Pedro issued an order that formally "forbade any meeting with López or one of his representatives . . . and decreed that any proposals of peace, or of an armistice, should immediately be rejected, no matter what the circumstances in which they were made."[44] Brazilian commanders consistently obeyed both the letter and the spirit of this command. Thus, in the negotiations at Yataity-Corá, the Brazilian general refused to shake López's hand and departed before any negotiations could begin. Likewise, several American offers of mediation foundered on the refusal of the Brazilians to contemplate negotiation.[45] This attitude was also in evidence after the fall of Asunción, which constituted the effective military denouement to the war: for Brazilian political leaders, including Emperor Pedro, as long as López remained in Paraguay the war was not over.[46] It was for this reason that the war could only end when the Brazilian army tracked down López's remaining force—by now consisting almost exclusively of the very young and the very old—and killed the Paraguayan leader on the field of battle. By this time, Paraguay had been utterly destroyed, while the costs of fighting the war had led to extensive internal rebellion in Argentina and had forced significant social changes in Brazil that eventually would lead to the overthrow of the monarchy.[47]

If Brazil had been committed to conquering and annexing Paraguay, this decision might have been understandable, although the war would still have been extremely costly. In reality, however, their actions after the war demonstrated quite clearly that the Brazilians had no intention of annexing or partitioning Paraguay. This situation thus raises a number of important puzzles. Why were the Brazilians so adamantly opposed to negotiation, especially when their ultimate political aims were in many respects quite limited? And why was their opposition to negotiation not shared by the Argentines?

In this section, I argue that unconditional surrender occurred in this case by the logic of the dispositional commitment problem: facing an unexpected, and unexpectedly ambitious, military attack, the Brazilians were unable to comprehend López's decisions and thus concluded that their adversary was implacably aggressive and that a sustainable peace would require regime change in Paraguay. This inference followed centrally from the disconnect between Paraguayan fears and Brazilian intentions: although López's belief that the Brazilians sought to extinguish Paraguayan independence was reasonable given the available evidence, it was incorrect. By contrast, available evidence indicates quite clearly that the Argentines did desire to annex Paraguay. In this context, however, they had a better understanding of López's decisions and thus did not reach the extreme conclusions of their allies.

Explaining a War to the Finish

Hypothesis 2b predicts that the target of a preventive war will be more likely to demand unconditional surrender when it does not intend to do what the initiator fears. López feared that his opponents were conspiring to extinguish Paraguayan independence and launched an aggressive war in an attempt to address that threat. Actions prior to, during, and after the war clearly indicate that his suspicions about Argentine intentions were largely correct, whereas his beliefs about the Brazilian intentions, if not unfounded, were untrue. There is good reason to believe that it was this difference that drove the variation in Argentine and Brazilian policies.

Consider first the evidence that the Argentines, or more accurately the *porteños* under Mitre, had never reconciled themselves to Paraguayan independence and would, given the opportunity, happily have extended their domain to Asunción and beyond. Under the Spanish Empire, Paraguay had been governed from Buenos Aires, and while internal disunion in the decades after independence militated against an attempt at restoring control, they retained a formal claim to Paraguay. Thus, for example, when Austria (at Brazilian behest) recognized Paraguayan independence in 1848, Juan Manuel de Rosas in Buenos Aires sent a long protest in which he asserted that "the Argentine Republic always preserved its rights over the territory of Paraguay and regards it as one of the

[101]

Argentine provinces."⁴⁸ In the negotiations over the terms of the Treaty of Triple Alliance, the Argentines were greatly frustrated by the (minimal) restrictions that the Brazilians wished to impose on Argentine demands, with Foreign Minister Rufino de Elizalde observing to the British representative in Buenos Aires that he "hoped he should live to see Bolivia, Paraguay, Uruguay, and the Argentine Republic united in one Confederation."⁴⁹ Once the fighting stopped, Argentine territorial demands were limited only by opposition of the Brazilians, who as the primary occupying power were however in a position to significantly limit Argentine territorial gains.⁵⁰ The Argentine leaders thus cannot have been overly surprised when López accused them of seeking to subvert his country's independence.

The discussion in the previous section demonstrated that López also had good reason to fear Brazil, given how little the Brazilians had done to allay Paraguayan concerns over the years. Yet with the benefit of hindsight we can see quite clearly that the Brazilians did not in fact intend to partition or annex Paraguay. The aforementioned negotiations over the Treaty of Triple Alliance provide one relevant data point: although the Brazilians have been criticized for their territorial promises to the Argentines, they held the line on maintaining Paraguay as an independent state.⁵¹ More important, once the war was over, the Brazilians were the primary occupying power, a result of their willingness to continue to fight after the Argentines had withdrawn all but a token force. In a position to impose whatever settlement they desired, they instead quickly signed a separate deal with the Paraguayans in which they were granted their relatively limited territorial demands and then backed Paraguay in forcing Argentina to accept less than had been promised under the Treaty of Triple Alliance.⁵² Under these circumstances, it is far from surprising that when López justified his war by claiming to be the target of Brazilian aggression, the Brazilians reacted with uncomprehending rage.

And it is not an overstatement to say that the Brazilians responded with rage. Facing López's invasion, the Brazilians discounted his stated concerns and convinced themselves that he must be some sort of miniature Napoleon, bent on continual expansion and hence a threat to peace as long as he remained in office. More specifically, they concluded, and publicly averred, that "they were fighting the unbridled ambitions of a tyrant and no one but they themselves could guarantee his extinction."⁵³ Given that inference, peace short of complete victory was folly: in short, in the assessment of one expert on the war, the Brazilians were motivated by an "ideological conviction . . . that López, and those like him, must go."⁵⁴ Subsequent historiography built on these wartime views, in part, as one scholar points out, because "[López's] own explanation for

[102]

his actions—that he was responding to aggression by Brazil which constituted a threat to Paraguay's survival—has not been taken seriously, on the ground that later history proves that Brazil was not the menace to his nation that Solano López imagined it to be"; this interpretation, as this scholar points out, is "hindsight with a vengeance," as it neglects the good reasons López had to suspect the Brazilians.[55]

Consistent with hypothesis 2c, this assessment contrasts sharply with the way the Brazilians had viewed López and Paraguay prior to the war. In the past, of course, the Brazilians had worked with Paraguay to check Argentina and had appreciated Paraguayan support for the overthrow of Rosas in 1851.[56] Prior to becoming president in 1862, López had been heavily involved in policy, for example mediating a peaceful solution to a dispute between Mitre and Urquiza in 1859 that brought him a brief period of general acclaim.[57] Before the Paraguayan declaration of war, few in Brazil had even heard of López, and the few who knew of him thought of him as the unimportant leader of an economically backward state, far from the monster he was eventually seen to be. The central role of the Brazilian inference that the cause of the war was the aggressive character of the Paraguayan leadership can also be seen in the disappearance of almost all animosity toward Paraguay once López and other high-ranking individuals had been replaced.[58]

The findings from this case thus strongly support the argument that "innocent" targets of preventive wars are more likely to demand unconditional surrender. In particular, the contrast between Brazil and Argentina is unusually illuminating: within the context of the same war, Brazil, which did not intend to do what Paraguay feared, demanded unconditional surrender, while the Argentines, whose motives were more in line with López's expectations, did not. Moreover, the trajectory of Brazilian beliefs accords with the expectations of the theory advanced here: it was only after his preventive attack that he came to be seen as an aggressive monster whose removal was a necessary precursor to a sustainable peace.

The Alternative Explanations

The three main rationalist explanations cannot provide a convincing account of unconditional surrender. While one could certainly attribute initial Brazilian intransigence to overoptimism, the difficulties that they encountered along the way to victory, and in particular the significant delay and escalation of costs following the disastrous Battle of Curupaity, provided ample opportunity for updating prior beliefs. That the Brazilians continued to refuse to negotiate beyond this point cannot be attributed to excessive optimism. There is also no great evidence that they were more optimistic about the ease of victory than were the Argentines.

Similarly, a domestic political story has significant weaknesses, except insofar as it acknowledges Emperor Pedro's ability to impose a policy that at least some in Brazil may have thought overly harsh. Of all the interested factions relevant for this conflict, the Unitarians in Argentina arguably benefited the most, as the country's unity was far greater after the war than it had been before, but Mitre was consistently open to negotiation.[59] By contrast, Emperor Pedro ultimately weakened his hold on power by insisting on the prosecution of the war to the finish, most notably by freeing slaves who were willing to fight and by opening the army—traditionally the prerogative of the nobility—to the lower and middle classes.[60] These changes turned the army into a significant force in domestic politics, one that was ultimately able to launch a coup in 1889 that overthrew Pedro and ushered in a federal republic.[61]

Nor can commitment problems grounded in a fear of decline provide a convincing account. The Brazilians appear to have recognized that the tide of history was on their side, and they certainly did not believe that Paraguay was headed for a significant expansion in capabilities, especially after the defeat of the forward Paraguayan army. The constant harping on López's iniquities contrasts with the absence of any commentary focusing on shifts in future capabilities. Certainly the refusal to consider the war over even following the fall of Asunción cannot be understood on the basis of fear of an imminent increase in Paraguayan capabilities.

Overall, then, this case provides strong support for my theory. Consistent with the argument, López initiated the war out of a preventive motivation, based on the fear that Paraguay's larger neighbors were conspiring to partition his country. The Argentines, who genuinely desired to annex Paraguay, understood these fears and thus treated López as a normal politician with whom negotiation was entirely possible. The Brazilians, by contrast, had no intention of taking over Paraguay; they misinterpreted López's behavior and concluded that only a complete victory could bring a sustained peace.

[4]

World War II

German Expansion and Allied Response

In contrast to the Paraguayan War and to many other bloody wars from history, which historians have often described as unnecessary and tragic, World War II seems less puzzling: Adolf Hitler did it. While ancillary questions remain, such as why the German people were willing to rally to Hitler's standard, the man himself has been seen as history's ultimate outlier, and World War II as simply the extension of his unique personality. It is likely for this reason, for example, that international relations scholars have spent far more time examining World War I, which is frequently seen as unnecessary and hence puzzling, than World War II. A serious attempt at explaining variation in the duration and severity of wars cannot afford, however, to simply paper over an extreme case as an unexplainable outlier, however attractive that approach may seem. It is thus worth asking to what extent the theory advanced here can account for German foreign policy prior to and during World War II.

Indeed, I argue that the preventive war mechanism provides an excellent account of Hitler's decisions. In this case, there are two separate components to Hitler's fear of decline. First, and more important, Hitler operated according to a clear and consistent ideology that informed him that absent significant territorial expansion Germany was headed for long-term decline. This belief thus was the basis for the *Lebensraum* policy and the conclusion that an expansionist war was in Germany's interest. Second, given that prior conclusion, Hitler then calculated that the late 1930s would be the optimal time to undertake that expansion, as rapid German rearmament and internal disorder in the Soviet Union had given Germany a temporary advantage that would not be sustainable in the long run. Thus, while the ideological basis for Hitler's actions informed him that an aggressive war to acquire territory in Eastern Europe was necessary, the material situation in Europe—in particular the anticipation of relative German decline over the next few years—explains why Hitler was so anxious to fight the war soon.

While Hitler's decisions guaranteed an unusually destructive war, they do not provide an entire explanation for the way the war ended. Even during the war's darkest hours, after the collapse of France and when the German advance into Russia seemed unstoppable, the Allies consistently rebuffed repeated attempts by a wide range of Nazi officials, military leaders, and others to reach some sort of understanding. By supporting unconditional surrender, Churchill, for all his greatness, effectively broke Britain as a world power, forcing the dissolution of the empire that he valued so highly and leaving his country permanently in the American shadow.[1] Soviet deaths numbered in the tens of millions, while the Americans availed themselves of a Soviet alliance whose costs they were to regret for forty-five years. Peace with Hitler, or with a military government that replaced him, would have been unpalatable, but almost all peace agreements contain something unpalatable for both sides, the more so when each has the ability to impose great costs on the other. That the Allies sincerely and consistently refused to negotiate with any possible interlocutor is thus, from the perspective of the standard bargaining model, simply astounding, and certainly worthy of further consideration.

The second half of this chapter thus analyzes the consistent Allied refusal to negotiate, focusing in particular on two periods—British obstinacy after the collapse of France and the Allied decision, announced in January 1943, to demand Germany's unconditional surrender—in which Allied leaders most explicitly discussed the question of whether or not to talk to Hitler or other German representatives. I first summarize the relevant history, in the process highlighting standard explanations for the decision not to negotiate. In each case, I argue that the unconditional surrender mechanism discussed in chapter 1 does a better job of explaining the refusal to negotiate than existing explanations or than any of the other three central mechanisms discussed in this book. Starting from the preventive motivations for German expansion, I argue that this case provides one of the clearest possible examples of innocence of intentions, as the Allies simply did not have the intentions that Hitler attributed to them. Lacking the correct understanding of Hitler's policies, the Allies concluded that he was engaged in expansion without limit, and thus that only fundamental change in Germany would bring a viable peace. Given this conclusion, the decision to fight to the end followed naturally.

[106]

The outline of German policy in World War II is well known, and thus the recapitulation here will serve primarily to identify the questions on which the rest of the analysis will focus. Prostrate after World War I, Germany began to reassert itself after Hitler's assumption of power in 1933, first breaching caps on the size of the German military and remilitarizing the Rhineland, and later resorting to aggressive and risky diplomacy to incorporate Austria and Czechoslovakia. The annexation of rump Czechoslovakia in early 1939, after Hitler had expressly promised that German territorial ambitions were sated at the Munich Conference the previous fall, spurred more active resistance from the European powers, and when Hitler turned to Poland war became inevitable. Just before crossing the brink, however, Hitler pulled off what was likely his greatest diplomatic coup with the Molotov-Ribbentrop nonaggression pact with the Soviet Union, which allowed Germany to avoid a two-front war.

The German army's series of stunning victories over the next year brought western Poland, Denmark, Norway, the Netherlands, Belgium, and France under Hitler's control, apparently freeing Germany from the danger of fighting a two-front war. At this point, Hitler considered the issue in the West to be decided and offered a settlement to Britain, which, while undefeated, had no plausible strategy to reverse the new status quo on the continent. Churchill's refusal left Hitler fighting a war that he did not want. His solution to this problem, however, has puzzled commentators ever since: unable to force a British capitulation, he launched an invasion of the Soviet Union. It was this decision, and to a lesser degree the full entry of the United States into the war following the Japanese attack on Pearl Harbor, that doomed Germany: rather than the quick victory in Russia that Hitler (and his Western opponents) expected, Germany found itself embroiled in a multiyear war against an opponent who would not capitulate despite crushing casualties and the loss of huge expanses of territory. The defeat at Stalingrad sealed Germany's fate, but Hitler showed no interest in negotiations with the Soviets, while his underlings tried to entice the Western powers into a separate peace and a crusade against communism. A convincing explanation for the duration and severity of World War II thus must confront Hitler's expansive political aims, his willingness to start one war while still fighting another, and his refusal to negotiate with the Soviets at the same time that he was open to negotiation with Britain.

To put the questions in a more positive form, why did Germany aim so high in World War II? Why was Hitler willing to undertake so many significant risks, including most obviously the decision to start new wars while old ones were not completed? And why was Hitler far more willing to negotiate with Britain than he was with the Soviet Union?

Historians have of course engaged these questions, albeit not from the theoretical perspective of the bargaining model. At one extreme are scholars, such as Hildebrand and Hillgruber, who argue that Hitler developed a blueprint for European domination, and potentially world conquest, even before seizing power in 1933, and that his subsequent actions merely implemented this blueprint.[2] At the opposite extreme are scholars such as Taylor, Broszat, and Mommsen, who contend that Hitler was fundamentally an opportunist primarily concerned with maintaining his position in Germany, and that the extreme policies that Nazi Germany adopted arose either from efforts to secure domestic legitimacy or from a bureaucratic race to the extremes in the absence of clear guidance.[3] This debate has obvious connections to the intentionalist-functionalist debate that focuses most heavily on the origins of the Holocaust, with intentionalists arguing that Hitler planned the Holocaust prior to seizing power, while functionalists argue that the Holocaust was an improvised response of the German bureaucracy to Nazi racial ideals and the absence of clear directives from above.[4] That said, explanations for Nazi aggression and for the Holocaust need not be the same, and my argument that Hitler intended to engage in significant territorial expansion (albeit not to the point of world conquest) prior to seizing power should not be seen as implying that he similarly entered office intending to carry out the Holocaust.[5]

As I discuss in greater detail below, I agree with the argument that Hitler planned to engage in significant territorial expansion at Russian expense even prior to seizing power, but disagree with strong arguments that he had mapped out the entire program prior to assuming power or that he foresaw either the conquest of or a final titanic clash with the United States after the acquisition of Lebensraum.[6] In the end, I argue that all of these questions can be answered once we understand both the long-term and the immediate decline that Hitler, guided by his ideology and his observation of the world, believed Germany to face. In this context, and given his beliefs about the intentions of his opponents, the extent and nature of German aggression becomes explicable.

Not Just Leaders, Not Just Information

Before examining these arguments, however, it is worthwhile to briefly review the limitations of informational and principal-agent answers to these central questions. With respect to optimism, while it is true that Hitler's military expectations were belied at times, in particular in the invasion of the Soviet Union, the decision to attack France was carried out despite low confidence in the probability of victory, while the reluctance to negotiate with the Soviet Union in the later stages of the war simply cannot be explained by a belief that future fighting would go well. A domestic political explanation is similarly unconvincing: while Hitler both worried about overthrow and enjoyed unusually wide freedom of action for a leader, the available evidence indicates that he sincerely believed that his policies would advance German interests and that much of the German public and military shared these beliefs.

To start with the informational mechanism, Hitler drew confidence in the likely outcome of a war from a number of sources, including the supposed martial superiority of the German race, an expected alliance with Britain, and a confidence in mechanized warfare and air power that led him to believe that World War II was not destined to be a repeat of World War I.[7] Prior to Operation Barbarossa, the confidence of both Hitler and his generals was buoyed by Germany's quick victories in the West and by Soviet weakness—as a consequence of the purge of the Red Army, roughly 75 percent of Soviet officers had less than a year of experience, creating limitations that were apparent in the Russo-Finnish War—and was consistent with the expectations of Germany's opponents in the West.[8] Advocates for the informational mechanism thus could point to high expectations prior to both invasions.

An informational perspective has difficulty accounting for the broader scope of Hitler's policies, however. He was confident from even before entering office that achieving his goals would require war; had he believed that his aims were justified by Germany's capabilities he logically should have expected that there would be at least some chance of achieving them at the bargaining table.[9] An informational perspective likewise has difficulty accounting for the major risks that Hitler was willing to run, as with the remilitarization of the Rhineland, which could have been achieved through negotiation had he been willing to wait but instead was undertaken unilaterally despite his knowledge that French intervention would have forced the Germans "to withdraw again with our tails between our legs."[10]

The most important decision of the war was to attack the Soviet Union while still at war with Britain, but the informational mechanism provides an essentially unconvincing explanation for that incident. While Hitler was quite confident that Germany would win, this confidence in itself provides no reason why Germany should start a second war before it had completed the first. The standard informational explanation for this decision was that Hitler had (incorrectly) concluded that the British were continuing to fight because they expected the Soviets to intervene and thus thought that crushing Russia would finally convince the British to capitulate.[11] Yet the Soviets were effectively allied to Germany and quite solicitous of German goodwill—had Hitler in fact believed that the British were holding out because they expected Soviet aid, a far more effective strategy would have been to keep the military pressure on Britain while continuing cooperation with the Soviets; invading the Soviet Union by contrast would involve giving Churchill exactly what he wanted.[12] This is not to say that Hitler did not hope that defeating the Soviets would finally convince the British to negotiate; rather it is to say that such a motivation provides at best a markedly incomplete account of the German decision.

Most important, the informational mechanism simply cannot explain the decision to refuse to negotiate on the Eastern front as Germany's military position deteriorated. While refusal to negotiate might be consistent with the informational mechanism in the opening stages of the war, when he might expect further victories that would give Germany more than Stalin would be willing to surrender at the bargaining table, after the Wehrmacht got bogged down outside Moscow, and especially after the catastrophe of Stalingrad, one would expect him to be interested in some sort of settlement with the Soviets as an alternative to a fight to the death that Germany looked increasingly likely to lose. Hitler clearly knew that the war was going poorly, but rather than scale down his demands he vested his hopes in the potential development of miracle weapons or the collapse of the opposing alliance. Similarly, the leading Nazis (with the possible exception of Goebbels) and generals harbored few doubts after Stalingrad about the eventual outcome of the war, but they too did not encourage negotiation with Stalin.[13] This behavior is simply inconsistent with the informational mechanism.

Domestic political answers to the central questions about German policy in World War II are similarly unconvincing. Hitler was admittedly an extreme autocrat with substantial policy freedom, although it is easy to overstate his autonomy, especially in foreign and military affairs where he had at times difficult interactions with the generals.[14] Throughout his time in office, he was particularly concerned about retaining the support of the general public and of the generals, and he tracked opinion in each group closely. Thus, for example, evidence that the German public was unenthusiastic about a possible war contributed to his decision to call off an invasion of Czechoslovakia in 1938.[15] Similarly, he was increasingly worried by indications of public dissatisfaction in the months prior to the invasion of France.[16]

The generals constituted the clearest constraint, however. Even after the stunning successes of 1940, Hitler complained that

> before I became Reich Chancellor, I believed that the General Staff was like a mastiff that had to be held firmly by the collar as otherwise it

would threaten to attack everyone. Since becoming Reich Chancellor I have been forced to the conclusion that the General Staff is anything but a mastiff. This General Staff has always tried to prevent me doing what I held to be necessary. . . . It is I who have always had to incite this mastiff.[17]

The most obvious manifestation of the degree to which Hitler had to worry about the generals' opinions is obviously the failed coup of July 1944, which came quite close to supplanting him. Although the failure of this coup led to the destruction of most of the German resistance, peace probes by Germans promising to overthrow Hitler in return for a negotiated peace continued until the end of the war, only to be stymied by the Allied policy of unconditional surrender.

A major component of Hitler's success was his ability to convince both the generals and a substantial portion of the German public that his policies advanced their interests. Thus, for example, a contemporary American journalist noted at the time that he had "still to find a German, even among those who don't like the regime, who sees anything wrong in the German destruction of Poland... . As long as the Germans are successful and do not have to pull in their best too much, this will not be an unpopular war."[18] A widespread fear of both communism and Russia and a sense of German superiority contributed greatly to a willingness to keep fighting on the Eastern front, even in the face of significant setbacks.[19] The high costs for getting on the wrong side of the Nazi state no doubt played a significant role in dampening opposition, but so did patriotism and a belief that Hitler was the best man to lead Germany through the crisis.[20]

Similarly, as Dale Copeland demonstrates, the generals mostly agreed with Hitler's goals; when they opposed him, it was because they were less optimistic about what the Wehrmacht was *capable* of achieving.[21] Once Germany's early victories validated Hitler's prognostications, the generals became quite supportive, however much they may have denied this fact when it became inconvenient after the war. Thus, for example, General Beck, who is best known to history for his leading role in the failed coup of 1944, concurred in a report that Germany "needs greater living space" that "can only be captured through a war."[22] The coup was driven by the realization that Germany was headed for inevitable defeat and that the Allies would not negotiate with Hitler: in this context, removing Hitler was a step to save the German state.[23] This discussion is not to say that support was unanimous, of course—many Germans were deeply and sincerely opposed to Hitler's policies. Enough supported him, however, that those who disagreed were unable to mount effective resistance.

[111]

Nazi Ideology and German Decline

A more convincing explanation for Hitler's policy choices must rely on his ideologically founded belief that Germany was headed for significant decline, especially relative to Russia.[24] As the argument here is unconventional and potentially controversial, it makes sense to lay it out in a reasonable amount of detail, starting with the history and core components of his ideology.[25]

Nazi ideology was not a fundamentally new creation, but instead drew on a number of currents in German thinking that had been around long before Weimar. While it contained some internal contradictions, especially as it pertained to domestic politics (where it was, for example, in Hitler's interest as a party politician to appear as both a capitalist and a socialist), and while its expression in *Mein Kampf* and in Hitler's unpublished second book was turgid and poorly organized, in the realm of foreign policy Nazi ideology contained a clear, stable, and basically coherent set of core ideas. Moreover, because intellectually Hitler was a reorganizer of existing ideas rather than a creator of new ones, his view of the world was readily understood by the soldiers, government officials, and regular Germans whose support he needed to prosecute the war.[26] Ultimately, it was this ideology, and the implications that it held for the German nation's position in the world, that made Hitler and the state he governed such a menace to the international system.

Hitler started from a desire for security for the German people; for him the issue was "nothing other than the question of preserving the German people's possibility of existence."[27] That existence, however, relied on the Germans' ability to protect themselves, both from external invasion and from the starvation that he predicted would soon beset a nation growing by 900,000 people a year while living in a territorially constrained space only one-fiftieth the size of Russia.[28] In short, absent the acquisition of significant additional territory (Lebensraum), the Germans would soon be forced, whether from voluntary population controls or simple starvation, to constrain their numbers while their neighbors continued to grow; when at last the world reached what he believed to be the inevitable stage at which total population matched the earth's potential output, the "brutal" and hence more populous races would be the ones to survive.[29] Indeed, Germany was uniquely constrained in this manner, as alone among the great powers it controlled relatively little territory both on its own continent and overseas in the form of colonies.[30] Thus, without additional territory, the Germans faced "the danger of vanishing from the earth or of serving others as a slave nation."[31] From the simple argument that territory determined population and that population and territory together determined security, Hitler thus derived a prediction that the

[112]

German nation was experiencing an extended and, if not addressed, terminal decline.

Solving this problem, however, would not be easy. Diplomacy might possibly restore territory lost in World War I (although he doubted it), but even a full restoration of the borders of 1914 would have been insufficient.[32] Moreover, he believed that any attempt at a diplomatic resolution to Germany's problem would inevitably fail given the racial enmity that Slavs and Jews supposedly held toward Germans: in his words, "You do not make pacts with anyone whose sole interest is the destruction of his partner."[33] Recourse to the most common strategy of dealing with decline—tacitly acquiescing in a loss of influence—was thus unacceptable. Instead, Germany needed to expand to new realms. This might be done through colonial expansion, but good colonies in which Europeans might settle were not available for the asking and pursuing colonial expansion would inevitably alienate Germany's otherwise natural ally, England.[34] If Germany had to fight to expand, far better that it do so in its own continent, where it could rely on its strength, land power. And he was convinced that the only reasonable target for such expansion was Russia, which simultaneously posed the greatest long-term threat and the best opportunity for overcoming that threat.[35] He rued what he saw as the missed opportunity of 1904, when German leaders might, in alliance with England, have taken advantage of Russia's defeat in war with Japan to address once and for all Germany's territorial problems.[36] The defeat of 1918 weakened Germany, but in one respect the war was beneficial, as it brought about the overthrow of the capable (and ethnically German) traditional rulers of Russia and replaced them with a group that, if bent on Germany's destruction, was far less capable.[37]

Thus already in the 1920s Hitler had developed the diagnosis of the German strategic situation that was to guide his policy upon his assumption of power. Germany was territorially constrained, and absent expansion it was doomed to population decay and eventual destruction. The constraints on the existing population and the loss (to racial miscegenation) of ethnic Germans outside the borders of the Reich meant that opportunities for addressing this problem diminished with every passing day.[38] Hence the urgency of quick action, supported by "undivided devotion" and the "harnessing of the very last possible ounce of energy."[39] He was not naive enough to believe that this aim could be achieved peacefully, but he argued that when it comes to self-preservation, "what is refused to amicable methods, it is up to the fist to take."[40] Indeed, far from shrinking from preventive war, he encouraged it. He expressed dismay at the failure to capitalize on the opportunity provided by the Russo-Japanese War, while hailing what he saw as Frederick the Great's preventive initiation of the Seven Years' War and taking issue with the claim that Bismarck opposed preventive action.[41]

That Hitler's ideology was wrong on many points does not mean that it was not sincerely believed. He truly considered himself to be a member of a race that was surrounded by implacable enemies and that was headed for decline, but that had an opportunity, if a fleeting and far from guaranteed one, to assure its own future security through aggressive action. While opportunistic in the nature of his subsequent expansion and more restrained in his public statements, he was consistently guided by the perceived need to acquire significant Lebensraum, as quickly as possible and at Russian expense. Thus, in his first meeting with the General Staff on assuming power in 1933, he laid out both the short-term aim of rebuilding the army and the long-term goal of "conquest of new living-space in the East and its ruthless Germanization."[42] Similarly, in a meeting with the chiefs of the armed forces in November 1937, he emphasized the importance of acquiring additional territory, revisiting and again rejecting alternate strategies to assure the nation's security, and declaring that failure to address this problem would ultimately pose "the greatest danger to the preservation of the German race."[43] One year later, at the height of the Sudeten Crisis, Hitler proclaimed himself entirely ready for war, "even if it lasts from two to eight years."[44] The connection between the belief in inexorable decline and the sheer ambition of his aims is entirely in keeping with the predictions of the preventive war mechanism.

Rearmament and the Window of Opportunity

Once in office, Hitler had an opportunity to move German foreign policy in what he viewed as the appropriate direction. He started by violating restrictions imposed by the Versailles Treaty on the size of the army and the militarization of the Rhineland. Next, taking advantage of the enshrinement of self-determination in Versailles, he began to push for limited territorial expansion as a first step on the road to acquiring sufficient Lebensraum. During the first few years of his rule, Hitler's foreign policy easily could have been that of a more conventional German nationalist, although given the urgency created by Germany's perceived decline he was willing to resort to risky faits accomplis rather than wait for diplomacy to yield its likely gains.[45] Behind the scenes, of course, his ambitions remained far greater.

By the late 1930s, these policies, in particular rearmament, had combined with Soviet military purges to give Germany a window of opportunity during which it had its best chance of winning the war for Lebensraum.[46] Already in 1937 he was forecasting that action must be undertaken by 1943–45, after which "only a change for the worse, from our point of view, could be expected."[47] The year before he had warned about the developing Red Army, observing, "One has only to compare the Red Army as it actually exists today with the assumptions of military men 10 or 15 years ago to realize the menacing extent of this development. Only consider the results of a further development over 10, 15 or 20 years and think what conditions will be like then!"[48] Indeed, between 1933 and 1939 the total number of men under arms in the Russian army more than quadrupled (to more than 4 million), while military spending went from 10 percent to 25 percent of the national budget.[49] These assessments conformed with those of the military, which broadly shared the fear of Russia's rise and believed it to be in Germany's interest to fight while still relatively strong.[50]

At the same time, the Germans were also constrained by the knowledge that France, and almost certainly Britain as well, would not simply accept the sort of expansion into Eastern Europe that Hitler envisaged.[51] The likelihood of British involvement, however, provided an additional reason for urgency, as both sides recognized that increased British rearmament meant that the German window of opportunity was rapidly closing.[52] In this context, Ribbentrop predicted that the British would play for time in diplomacy, while Germany should move quickly while its advantages persisted. This attitude carried over once war was declared: in the closing days of the defeat of Poland, Hitler was already pushing for an invasion of France *that fall*, on the grounds that "time may be reckoned more probably as an ally of the Western Powers rather than of ours," while the British and French were content to hide behind the Maginot Line—the main French fortifications—in the belief that rearmament and economic warfare were continually sapping Germany's ability to fight.[53]

Germany's temporary advantage was further heightened by Stalin's purge of the Red Army officer corps in 1937 and 1938, which left the Soviet Union unprepared for war in the immediate future, as the poor performance in the Russo-Finnish War was to demonstrate.[54] Hitler was fully aware of the effects of the purge, which he found inexplicable but which afforded him greater freedom of action, at least in the immediate term.[55] Most important, Hitler recognized that Stalin's desire to avoid a war while the Red Army recovered would make him amenable to a nonaggression pact that would allow Germany to avoid having to fight a two-front war.[56] Thus, after the fall of Poland, Hitler observed that "at the present time the Russian army is of little worth. For the next one to two years the present situation will remain."[57] Indeed, Molotov reported that Stalin "had felt that only by 1943 could we meet the Germans on an equal footing," thus militating against Soviet intervention in the western war.[58] Despite what he saw as the fundamental antagonism between the

Germans and the ethnically Slavic and ideologically communist Soviets, Hitler was quite confident of Soviet neutrality once the war began, at least in the short term, precisely because the Soviets stood to gain from delay. The opportunity to dispatch France at a time when the Russians would prefer not to fight, as well as the subsequent opportunity to attack the Soviet Union before it had recovered from the purge, thus provided an additional incentive for haste.

The deal with Stalin was always intended to be temporary, however, and Hitler was eager to make it as temporary as possible. Even while the war in the West was still raging, he remarked to General Jodl that he would "take action against this menace of the Soviet Union the moment our military position makes it at all possible"; Jodl himself subsequently commented that "it was better therefore to have this campaign now, when we were at the height of our military power."[59] In December 1940 Hitler remarked that Germany needed to deal with Russia in the next year, "because in 1942 the United States would be ready to intervene."[60] The same month, he gave the formal order to begin planning for Operation Barbarossa, the massive invasion of the Soviet Union initially scheduled for the following May. In internal planning, the preventive nature of this invasion was anything but a secret; indeed, Hitler explicitly referred to the war as preventive on June 6, 1941, only a few weeks before the attack.[61]

Some have argued, based on Hitler's apparent belief that the continuation of the war in the West was a consequence of "Britain's hope on Russia," that Germany invaded Russia to convince Britain to come to terms, and that from this perspective Barbarossa was Hitler's greatest mistake.[62] The evidence presented here indicates the problems with this perspective, however. The desire for a war with Russia was present in *Mein Kampf*, when Hitler believed that Britain would be on Germany's side. That Hitler was shifting his sights eastward as soon as the defeat of France became apparent, and before the extent of British obstinacy had become clear, likewise demonstrates that the invasion of the Soviet Union was always a primary motivation. In sum, if Russia's defeat brought Britain to terms, that would be a bonus, but the war with Britain was always a sideshow to the main operation. Indeed, given that the war in the West was undertaken to clear the path for the war in the East, and given the belief that any delay weakened Germany relative to its opponents, from within Hitler's view of the world the decision to attack in the summer of 1941 was the correct one, even if the invasion ultimately miscarried.

Overall, the history of German foreign policy prior to and during the war conforms strongly with hypothesis 1b, which predicts that leaders who fear decline will adopt unusually risky policies. As was previously noted, Hitler's diplomacy in the 1930s repeatedly achieved by fait accompli—with the attendant dangers of armed retaliation by France or other neighbors—what a more patient leader might easily (if less quickly) have achieved through more patient negotiation. At Munich, he came close to throwing away an agreement that neutered Czechoslovakia as an effective opponent out of a desire not to postpone the inevitable war. Although far from obvious in hindsight, the blitzkrieg invasion of France also carried serious risks to the elite German units at the core of the invasion force. On the basis of war games prior to the invasion, German generals concluded that victory relied almost entirely on the complete success of attempts to hide the German drive through the Ardennes; had the British and French caught on more quickly and called back units sent north into Belgium and the Netherlands, the invasion of 1939, like that of 1914, might have failed to bring the decision that the Germans needed.[63] Finally, and most obviously, he launched the invasion of the Soviet Union without first defeating Britain.

To summarize, while the ideological belief in the decline of the German nation lay behind Hitler's quest for Lebensraum, more immediate calculations of shifting capabilities helped determine the timing and direction of his expansion. Indeed, as late as 1937, he hoped to have another six years for preparation, but quick British rearmament combined with the window of opportunity created by the Soviet purge provided reason to accelerate the schedule.[64] The belief that time favored Germany's opponents, which suffused military and political discussions prior to and during the war, provided a reason for urgency and helped motivate Hitler's war aims as well as the risks that he was willing to run in their pursuit.

Negotiation versus War to the Death

The rationale behind Hitler's expansive political aims and risky foreign policy prior to the war helps to explain otherwise puzzling decisions taken once the war was underway. In addition to his willingness to take risks—evident most obviously in the decision to invade the Soviet Union before Britain was defeated—the belief that Germany faced terminal decline relative to Russia helps to explain why Hitler repeatedly pursued negotiations with the British while showing no interest in a settlement with Stalin.[65] Starting basically from the British and French declarations of war, Hitler and those under him sought to restore peace with the Western powers, especially with Britain. Almost immediately after the collapse of Poland, Hitler made a public appeal for peace on the basis of a free hand for Germany in Eastern Europe and the restoration of some of Germany's lost colonies.[66] During the same period, Hermann Göring sought more quietly behind the scenes to establish peace with the

British.[67] Indeed, in his discussions with his generals Hitler repeatedly indicated his confidence that the British would come around, at various points in May and June of 1940 observing that his goal was "to arrive at an understanding with Britain on the basis of a division of the world," that the British could have "a special peace at any time," and that "his aim was to make peace with Britain on a basis that would be consistent with her honor to accept."[68] Ultimately, he only abandoned the pursuit of peace with Britain once the British made it unambiguously clear that they would not countenance settlement, and even then his underlings, especially Himmler, continued to pursue back-channel possibilities.[69]

By contrast, evidence of attempts to reach peace with the Soviets once the war began is slim and unconvincing. There is speculative evidence that the Soviets may have sent peace feelers via Bulgaria in October 1941, but this evidence is uncertain, and if such feelers were sent they were definitely rejected.[70] Others have pointed to rumors of approaches in 1943 and 1944 by the German Peter Kleist to Soviet representatives in Stockholm, but these approaches, if they happened, did not have official authorization and achieved nothing.[71] Most important, whereas we have clear and unambiguous evidence of Hitler's openness to negotiations and an eventual settlement with Britain, there is no similar evidence of plans for negotiation with the Soviets. Similarly, whereas he clearly was willing to leave the British Empire alone and appears to have envisioned the survival of France (minus "historically German" territory along the Rhine), German plans for the East involved the complete destruction of Russia in Europe, including the razing of Leningrad and the incorporation of Poland, the Baltic region, and Ukraine into the Reich, with only a rump Russia surviving behind the Urals.[72]

This contrast is hard to explain from many perspectives but is completely consistent with the shifting power motivation for war. Given the belief that Germany's primary problem was eventual decline with respect to Soviet Russia, war with Britain was a regrettable distraction, which furthermore the British, if they understood Hitler's motivations, would recognize was unnecessary—German expansion in Eastern Europe would not undermine the British Empire. A negotiated settlement with Britain was both desirable and, he expected, achievable. By contrast, the fight with the Soviet Union was the one that he actually wanted, and he believed that victory could only be achieved via the destruction of Russia in Europe. Given that goal, settlement on intermediate terms was unacceptable, and thus discussions of such settlements (which of course were the only ones that the Soviets would countenance so long as they had an army in the field) were simply a waste of time. Similarly, Hitler's aims on the Eastern Front did not wax or wane with German victories or defeats, instead remaining remarkably constant from the writing of *Mein Kampf* through to the final Battle of Berlin.[73] Given the belief that a Germany that failed to achieve Lebensraum was doomed anyway, he saw no reason to reduce his war aims as the battles turned against Germany, instead putting his faith in increasingly unlikely scenarios that might bring about a dramatic military reversal and eventual victory.

While Hitler was not entirely insensitive to costs—at several points in his writings he decried the extreme waste of German blood in World War I in pursuit of what he saw as an inappropriate aim, for example—given what he saw as the costs of inaction he was willing to countenance German (and Allied) deaths on an almost unimaginable scale.[74] A leader who could not be deterred by even extraordinarily high costs of war and who refused to moderate his (always extravagant) demands in response to a steadily deteriorating military position guaranteed that no peace settlement would be possible and thus that the war on the Eastern Front—which lasted almost four years and in which tens of millions of people died—would continue until the total defeat of either Germany or Russia. The connection between the belief in German decline, the desire for significant gains at Russia's expense, and the refusal to negotiate with Stalin once the war began, conforms closely to theoretical expectations and provides strong support for hypothesis 1, which predicts that wars driven by preventive motivations will be tend to be unusually long and deadly.

From a theoretical perspective, one might wonder why Germany did not capitulate as defeat became increasingly inevitable. After all, theory predicts that once a feared decline has occurred and the opponent is in a position to impose today the concessions that were expected down the line, there is no reason to continue to fight. The multilateral nature of the war, however, militated against such a development: down to the final days, the Germans could hope that the opposing coalition might splinter and the Western powers, whose concerns about the Soviet Union were well known, might come to Germany's aid.[75] This hope provided reason not to seek terms with Stalin, especially as the Soviets had not signaled an interest in a negotiated peace. Furthermore, anyone who sincerely believed Hitler's racial theories expected that the consequences of defeat would be the extermination of the German nation. Given this expectation, acknowledgement of defeat as long as there was any possibility of a recovery was simply suicide.

The Limits to German Expansionism

The discussion thus far has made the argument that a belief in German racial decline motivated an ambitious and risky attempt to expand into

Eastern Europe, an attempt that ultimately failed but that came remarkably close to success. Some readers, however, may object that the perspective on Hitler advanced here does not conform to their understanding of the man, in particular with respect to his final political aims. In the academic discussion of international relations, Hitler's Germany is the paragon of the "unlimited aims revisionist," a country that would continue to expand until either it was stopped or it had conquered the world.[76] My argument, however, implies that Hitler's aims, while certainly not small, also were not unrestricted.[77] As the belief that Hitler's aims were in fact unlimited is widely held, and as such aims would be inconsistent with the argument advanced here, it is worth briefly examining the evidence that has been advanced in support of this claim.[78]

Historians have come down on both sides of the question about the extent of German war aims, with the distinction clearest in a debate between so-called globalists and continentalists.[79] Globalists traditionally have supported their position with respect to three categories of evidence: various comments attributed to Hitler over his time in office, in particular the comments recorded by Hermann Rauschning; available evidence about long-term German military planning; and Hitler's discussion of the United States in his unpublished second book. On closer examination, however, all three of these lines of evidence are either of questionable validity or amenable to an alternate and more convincing interpretation.

The simplest way to support the argument that Hitler sought to conquer the world is to point to instances in which he reported such a desire himself. To this end, Hermann Rauschning's recollections of personal conversations with Hitler provide the most useful evidence: in more than one hundred reported conversations, all from his early years in power, Hitler speaks openly of his intention to conquer the world and lays out plans that include the conquest of Africa and the Americas.[80] Subsequent historians, however, have concluded that the book is fabricated, more a work of wartime propaganda than a useful source on Hitler's thinking.[81] Outside this discredited source, other evidence of Hitler's stated intention to conquer the world is far more circumstantial, consisting of one-off comments, such as the ambiguous references to Germany becoming a "world power" (*Weltmacht*) in *Mein Kampf*.[82] Overall, in the words of one historian, "no evidence exists setting forth Hitler's declared intention to conquer the world."[83]

A second line of argument has focused on German military plans, and in particular plans (never realized) to build a significant surface fleet as well as potential plans to build long-range bombers to be stationed on the Azores, where they could strike at the United States.[84] Here again the evidence is far from convincing, however. The cited military plans were

[120]

never really pursued, and in any event are amenable to less expansionist interpretations. Thus, for example, plans for naval development were designed to address European problems—the blockade of Britain or a possible falling out with Spain.[85] Similarly, the desire to develop the capacity to bomb the American mainland was a reasonable response to the strategic problems created by the anticipated entry into the war of an opponent whom Germany could not directly attack: the Germans believed that an ability to impose direct costs would increase the probability that the Americans would back down, just as they hoped that bombing London would induce Britain to agree to a settlement.

The final argument for unlimited aims has focused on apparent evidence that Hitler foresaw a final titanic clash with the United States at some point, perhaps after his death.[86] This argument relies primarily on the apparently dramatic revision in Hitler's thinking about the United States between the time that he wrote *Mein Kampf* and the time that he wrote his unpublished second book in 1928, in which he went from describing the United States as weak and irrelevant to describing it in far more positive terms and raising it as a possible future opponent.[87] This discussion, however, occurs in the context of an analysis of a proposal for a pan-European policy; in arguing against this policy, Hitler found it useful to have an example of a nation that expanded by conquest and then populated the conquered lands itself—as he argues Germany should do in Europe—and thus presents the United States in a light that is consistent with this argument. In short, the United States is not even the subject of this discussion and enters in a way that is best suited to serve Hitler's rhetorical purpose.[88] Consistent with this view, Hitler reverted subsequently to his standard dismissals of the United States thereafter.[89] Similarly, Hitler's oft-cited prediction of a final titanic clash with the United States was made to Soviet foreign minister Molotov at a time when he was trying to convince the Soviets to accept a manifestly unequal deal that would allow Germany to dominate Europe in return for Russian expansion in Central Asia; in this context, an appeal to an external enemy was rhetorically useful.[90]

Ultimately, the claim that Nazi Germany was an unlimited aims revisionist is, I would argue, a theoretical one grounded in the observation that every German gain under Hitler was simply the prelude to another attempt at expansion. In short, given a leader who repeatedly lied about the extent of his aims, claiming that each territorial demand was his last, why should we believe that the land grab that failed would have proved the final act of expansion? Absent a good explanation for why Hitler might seriously believe that he needed to take over most of Eastern Europe but not the rest of the world, it is not unreasonable to believe that he in fact sought to conquer the world. In contrast, the observation that from the writing of *Mein Kampf* he identified Lebensraum acquired from Russia as the primary goal, that evidence of plans for further expansion is at best quite weak, and most important, that a drive to acquire Russia's territory in Europe (but not more) was the logical implication of his theory of the world all provide evidence that German expansion would indeed have stopped of its own accord.

UNCONDITIONAL SURRENDER AND WAR TO THE DEATH

German aggression is, however, only part of the puzzle of World War II. The remainder of this chapter examines the Allied refusal to negotiate, focusing in particular on the British and the Americans, with whom Hitler was quite interested in reaching an agreement. Although typically seen as self-explanatory, the Allied refusal to negotiate is deeply puzzling from the perspective of the bargaining model, given the extent of German military successes and the high cost of conquering Germany. By examining both the British refusal to negotiate in the summer of 1940, for Britain the war's darkest hour, and the joint Allied decision to demand Germany's unconditional surrender, I argue that this behavior can only be explained by reference to the dispositional commitment problem: the Allies, who did not share Hitler's theory of international politics and hence did not understand his motivations for expansion, concluded that the Germans were committed to aggression, and that peace could be ensured only by fundamental reform of German government and society.

British Intransigence in the Absence of Options

From the British and French declaration of war in September 1939, Hitler was already publicly discussing settlements to restore peace.[91] His opponents, however, refused to negotiate. During the Phoney War, the British and French rebuffed both these proposals and several offers of mediation from neutral powers while developing plans for a multiyear war that would ultimately wear Germany into submission, as they had done in World War I.[92] The British expected an initial German attack that would break against the Maginot Line, after which the combination of trench warfare and economic blockade would undermine the German economy and eventually force Hitler out of power.[93] This strategy was critically dependent on the French: General Edmund Ironside in January 1940 wrote in his diary that "we must have confidence in the French army. It is the only thing in which we can have confidence. Our own army is just a little one and we are dependent upon the French. . . . All depends on the French army and we can do nothing about it."[94]

[122]

The catastrophic defeat of the French army in May and June 1940 thus eliminated what Lord Halifax, the British foreign minister, described as "the one firm rock on which everybody had been willing to build for the last two years," leaving Britain's strategy for victory in tatters.[95] With the French gone and the Low Countries in German hands, the British not only lost the primary means of striking at Germany; they also, for the first time since 1805, had to confront the imminent threat of an invasion of the British Isles. Reports commissioned once the possibility of French collapse arose paint a dark picture: to continue the war, Britain would have to resort to "a form of government that approached the totalitarian," and even then the best hope was simply to hold out until the United States entered the war, which was possible only so long as the RAF survived.[96] Moreover, the British overestimated the impact that German aerial bombardment would have—reasonable people (including Prime Minister Churchill) expected that the first week of bombing would kill thirty thousand people in London alone—providing a good reason to seek peace earlier rather than later.[97]

Moreover, despite this catastrophic defeat, Hitler's proposed terms of peace with Britain were quite moderate: rather than bearing, as Churchill predicted, a particular animus toward Britain, he offered to leave the British Empire untouched, even withdrawing prior demands for Germany's old colonies.[98] In a series of intense discussions in late May 1940, at arguably the darkest point of the entire war from the British perspective, the War Cabinet debated whether to join the French in accepting Italian mediation, ultimately deciding against this and any other form of negotiation.[99] Scholars have subsequently mined these discussions quite thoroughly, although debates remain about the extent to which any of the principals directly advocated accepting Hitler's terms. Viscount Halifax, the foreign minister and the strongest proponent of agreeing to the French approach, justified his position by arguing that rejecting the French proposal would make them more likely to capitulate in the immediate future, rather than on the grounds that Britain should negotiate with Germany.[100] That said, the possibility of negotiation was definitely raised at several points, and Churchill won the debate largely by insisting that negotiation with Hitler was generally unacceptable. The content of their discussions thus provides insight into why, despite their military defeat, the British rejected the French proposal to seek mediation and turned down Hitler's subsequent proposal of peace on the basis of the military situation at the time. Although the invasion threat lasted through the summer and the U-Boat attacks on British shipping constituted a serious threat through the following spring, the War Cabinet never again seriously contemplated an approach of any form to Germany.

[123]

War to the Death: Unconditional Surrender

In the summer of 1940, the British sought only to survive; by the end of 1942 it had become more reasonable to talk publicly about the terms of a possible Allied victory. At the Casablanca Conference in January 1943, Churchill and Roosevelt duly announced their conclusion that "peace could come to the world only through . . . the unconditional surrender of Germany, Italy, and Japan."[101] In practice, however, this statement merely constituted the formal announcement of a decision that had been made substantially earlier.[102] The Americans had begun internal planning for the postwar world almost immediately after Pearl Harbor, most notably through the establishment of the Advisory Committee on Post-War Foreign Policy.[103] The committee dealt with a wide range of issues related to war termination and the postwar peace, focusing in particular on the nature of the international organization that would be created once the war was won. The discussion of war termination, by contrast, was remarkably brief, as from the outset the participants unanimously agreed that peace could only come through the total defeat of Germany and Japan. Thus, for example, on March 21, 1941, the committee met to consider the sequence of events that would follow on war termination through the surrender of the enemy; no other possible way of ending the fighting was considered.[104] The discussions of the political subcommittee "throughout were founded upon belief in the unqualified victory by the United Nations," and the security subcommittee rapidly concluded that "as between a negotiated cessation of hostilities or armistice on one hand and an imposed unconditional surrender on the other . . . nothing short of unconditional surrender by the principal enemies, Germany and Japan, could be accepted, though negotiation might be possible in the case of Italy."[105] Nor was this stance limited to participants in the Advisory Committee: in September 1941 the Chiefs of Staff, General Marshall and Admiral Stark, submitted a report on production requirements in which they conveyed their conviction that

> the first major objective of the United States and its Associates ought to be the complete military defeat of Germany. . . . An inconclusive peace between Germany and her present active military enemies would be likely to give Germany an opportunity to reorganize continental Europe and to replenish her strength. Even though the British Commonwealth and Russia were completely defeated, there would be important reasons for the United States to continue the war against Germany, in spite of the greatly increased difficulty of attaining final victory.[106]

This statement—three months *before* Pearl Harbor and the formal American entry into the war—demonstrates the degree to which key American policymakers had concluded that Germany *by nature* posed a threat that could only be eliminated by total military defeat and the remaking of the German political system. Roosevelt's statements betray a similar conviction: thus, for example, in March 1941 he pledged maximal American aid to the opponents of the Axis powers "until total victory is won," and in a December 1940 Fireside Chat he averred that "there is far less chance of the United States getting into war, if we do all we can now to support the nations defending themselves against attack by the Axis than if we acquiesce in their defeat, submit tamely to an Axis victory, and wait our turn to be the object of attack in another war later on."[107] When Roosevelt announced at Casablanca that the Allies would only accept unconditional surrender, therefore, he was advancing a long-established position that accorded with the views of a wide range of Americans.

Nor did the commitment to unconditional surrender waver subsequently. German peace feelers grew increasingly importunate as the war turned against the Axis powers—Heinrich Himmler in particular proposed a variety of deals, including an alliance against the Soviet Union or the ransom of Jews set to die in the Holocaust.[108] In every case, however, the Allies responded that no peace on terms other than Germany's unconditional surrender was acceptable, even when the Germans accepted that total occupation was inevitable and sought only to make sure that the British and Americans would be the primary occupying powers.[109] Roosevelt also insisted that the surrenders of Germany's allies be officially unconditional, although (as discussed in more detail below) the Allies in practice were open to fudging the content of unconditional surrender in a way that they were not for Germany or (until the very end) Japan.

The Standard Explanations and Their Limits

Scholars have proposed a number of explanations for both Britain's refusal to negotiate and the decision to demand unconditional surrender. None, however, provide a wholly convincing account of what in comparative terms are very unusual policy decisions. This section reviews those arguments, connecting them where relevant to the informational, principal-agent, and situational commitment problem mechanisms.

Hopes, Not Expectations

In the context of catastrophic defeat, the informational mechanism clearly predicts that the British would have lowered their political demands in an effort to escape from an increasingly unattractive war. That they not only rejected Hitler's proposals but rejected the entire possibility of negotiation thus would seem to be prima facie evidence that the informational mechanism played little role in this case. Readers may be surprised to learn, then, that the standard explanation for British policy is basically consistent with the informational mechanism. According to this view, which is most closely associated with the historian David Reynolds, the British fought on because they sincerely (if incorrectly) believed that American intervention and attacks on the German economy would quickly bring the Germans to their knees.[110] Such optimism might indeed justify rejecting negotiation at a time when Hitler believed himself to be in a position to dictate peace terms. A closer look, however, identifies some significant problems with this argument. At the core, the central problem is that British leaders, simply put, were not particularly optimistic about Britain's prospects.

Consider first the possibility that British leaders believed that the United States would intervene and turn the tide of the war. It certainly is true that Churchill pushed the Americans for every ounce of support they could provide, even ending his famous "we shall fight" speech by looking forward to the day when "the New World . . . steps forth to the rescue and liberation of the old."[111] That said, he had few illusions about the willingness and ability of the Americans to provide practical assistance in the immediate future: antiwar public opinion meant that Roosevelt could not declare war, and even had he been able to overcome that obstacle the immediate effect of American entry would have been minimal given the time it would take to remobilize the American army, which had been all but disbanded after World War I, and to redirect the economy for war production. The British leadership was fully apprised of this situation—indeed, for Churchill, the primary benefit of American intervention would be the fillip it would give to British (and, prior to their capitulation, French) morale.[112] While American intervention did ultimately help Britain to win a long war, in the summer of 1940, when the British feared that a German invasion force might arrive any day, American entry would have done little to help Britain survive. To argue that Churchill and the War Cabinet rejected negotiation because they expected that American intervention would dramatically improve their military prospects in the short term thus is simply untenable.

A slightly stronger case can be made for the argument that British leaders resisted settlement because they believed the German economy to be vulnerable. Based on a reading of World War I that saw the blockade of Germany as the primary determinant of victory, the British committed to a plan of using blockade and aerial bombardment to weaken the German economy and thereby secure victory.[113] The belief that attacks on the German economy might prove fruitful endured even through the collapse of the French army. Thus, for example, in a War Cabinet meeting on May 26, Chamberlain averred that Hitler would

[126]

need to win by winter, while Halifax used the putative German economic weakness to cast doubt on Churchill's claims that Hitler would advance maximalist demands.[114] Churchill, always the strongest proponent of continuing the fight, also held to this view, arguing on June 13 that the blockade would become increasingly effective.[115]

With the defeat of France, however, the British were incapable of seriously targeting the German economy. The blockade was already badly weakened by Soviet shipments under the Molotov-Ribbentrop Pact, while the need to take over French duties fighting Italy in the Mediterranean left the Royal Navy badly overstretched. The fallback plan of resorting to aerial bombardment was no better.[116] Initially reluctant to resort to bombardment because of fear of German retaliation in kind, the British ultimately discovered that aerial bombing was far less effective than advertised: a wartime study found that only one third of British planes managed to drop their payloads within *five miles* of their target.[117] Given the high costs, especially of daylight raids and raids beyond the range of fighter escorts, the British were rapidly forced to acknowledge that a knockout blow "was utterly beyond Bomber Command's capacity or means in 1940" and thus to curtail aerial bombardment.[118] Realistically, therefore, the strategy of economic coercion could be expected to work, if at all, only over a long time frame—indeed, given that the British initially expected that it would take several years for Germany to collapse, it would be surprising for them to expect the strategy to work better once it became clear that Germany would not have to fight an extended war of attrition along with border with France. When we combine these difficulties with the immediate threat of a German invasion and the longer-term threat from the German U-Boat blockade of Britain (which would come to constitute a serious threat to Britain's ability to stay in the war), it was entirely possible that Britain would be defeated before the attacks on the German economy even began to take effect.[119]

Indeed, whatever Churchill's public confidence in ultimate victory, in private he and Britain's other leaders were far more pessimistic. Once the scale of the disaster on the continent became clear, Churchill asked of his General Staff only whether they could "hold out *reasonable hope* of preventing serious invasion," while Halifax acknowledged that "it was not so much a question of imposing a complete defeat upon Germany but of safeguarding the independence of our Empire and if possible that of France."[120] Similarly, in letters to Roosevelt, Churchill repeatedly raised the specter that a future government might capitulate—accepting a status as a Nazi protectorate and surrendering the fleet—in return for peace.[121] To take one historian's assessment, the policy of continuing to fight "did carry with it the possibility of ultimate victory should Hitler blunder, but also the more likely spectre of national annihilation."[122]

[127]

Ultimately, British confidence in the summer of 1940 can be better interpreted as an example of motivated bias than as the informational mechanism at work.[123] One of the motivations behind Chamberlain's appeasement policy was the recognition that Britain could neither win a short war nor afford a long one; that basic constraint did not disappear once war actually arrived, and thus British leaders went searching for a scenario in which they might achieve victory in short order without actually having to do much fighting.[124] Having adopted a strategy of indirect attack, they continued to hold out great hopes for this approach until Soviet and American involvement supplied them with a strategy that was actually viable. In an assessment after the fact, one commentator noted that

> so long as the enemy held the initiative, and especially after the collapse of France and while American opinion was resolute not to enter the war, there was bound to be something unrealistic about many appreciations and proposals. The writers of course assumed their country's survival and . . . eventual victory. But how that victory was to be won could not be forecast. What was required was not detailed forecasts of the future but practical recommendations as to how to keep our heads above water through the critical months immediately ahead and how to preserve a correct balance in our plans for expansion. This should be remembered if some of the appreciations of the early phases of the war seem unduly optimistic.[125]

In other words, faced with a difficult situation from which they saw no escape, the British conjured up the best plans that they could and then put their hopes in them.[126] Of course, a motivated bias requires a motivation: if a conviction that peace was impossible led British leaders to conclude that they might still be able to win, the true explanation for the decision to fight lies in that prior conviction that peace was impossible. The informational mechanism provides no explanation for that belief.

Unconditional Surrender

In contrast to the plausible, if ultimately unconvincing, argument that the informational mechanism drove the British refusal to negotiate in the summer of 1940, there simply is no plausible informational interpretation of unconditional surrender. Under the informational mechanism, unconditional surrender would only be a viable policy if the side demanding it expected to be able to achieve victory both with a probability approaching certainty and, equally importantly, at quite low cost. By contrast, in World War II the Allies were confronted with an opponent with the most capable army in world history that at the time of the Casablanca announcement controlled territory stretching from the Atlantic almost to the Caspian. Even after the Soviet victory at Stalingrad—the turning point of the war—the Allies could not be entirely confident of victory, and no one could deny the high costs that the complete reduction of German forces would require. Given these costs, an informational approach would predict that the Allies would be open to negotiation on some terms (if not necessarily terms to which Hitler would agree); categorical refusal to negotiate is simply inexplicable within this mechanism.

Domestic-Political Unity

A domestic political (i.e., principal-agent) explanation for these decisions would highlight private ends that the Allied leaders might achieve through continued war, which they would attempt to achieve by misleading their publics about the likely course of future fighting. While one might argue that the Allied leaders benefited personally from a continuation of the war, it is impossible to argue that they attempted to mislead their publics about the difficulty of achieving military victory. That their publics remained supportive of the war effort indicates that they generally believed a continuation of the war to a decisive finish to be in the national interest.

It is true that both Churchill and Roosevelt arguably stood to gain from the continuation of the war. Churchill was an unpopular iconoclast who never would have gained power absent the war and who did not foresee much of a future for himself as prime minister once the war was over.[127] Moreover, he was dependent on the support of figures like Chamberlain and Halifax who were more highly esteemed than he when he entered power.[128] Roosevelt likewise benefited personally from the war, as the unsettled international political situation provided a justification for violating the longstanding norm against presidents serving more than two terms in office. Moreover, convinced of the danger posed by Germany and Japan, he was willing to mislead the American public about the nature of his commitments in an attempt to provide the maximum possible aid to Britain prior to Pearl Harbor.[129]

Once their countries were at war, however, neither leader engaged in the deception that is central to the principal-agent mechanism; indeed, both took pains to guard against unwarranted optimism. Thus, after the unexpected successful evacuation of the British Expeditionary Force at Dunkirk, Churchill reminded the public that Britain had nonetheless just suffered a "colossal military disaster" and that "wars are not won by evacuations."[130] The American leadership likewise moved away from misrepresentation once in the war. Roosevelt in fact toned down his rhetoric on total victory after Pearl Harbor because he feared that it might lead to public overoptimism about the ease with which such a victory could be achieved; he also repeatedly emphasized the threat posed by Nazi Germany to the United States.[131] Similarly, the day after the attempted assassination of Hitler, Secretary of State Hull commented that the attempted coup indicated that the Germans recognized their impending defeat but cautioned that "we should not let these apparent developments give rise to over-optimism," for "the fighting ahead will be hard."[132]

Given the absence of significant deception, as well as the high salience of the war, the principal-agent mechanism would predict that leaders who were pursuing personal ends would face increasing criticism over time, ultimately forcing them to settle. Instead, both the public and political opponents rallied behind the leadership in Britain and the United States. This support was crucial—Churchill, for example, was initially extremely reliant on the support of Chamberlain and Halifax, and was able to get his way only because the two (especially Chamberlain) backed him.[133] Likewise, the Labor Party consistently backed Churchill both in the War Cabinet (where it had two representatives) and in public, despite the recognition that the war would undermine its cherished social program.[134] Public opinion likewise backed the decision to continue to fight, even while recognizing that the fall of France was a calamity.[135]

The story was similar in the United States, where both elite and mass opinion rallied behind Roosevelt after the Pearl Harbor attack. Whereas 37 senators and 194 members of the House of Representatives had voted in November against a revision of the Neutrality Acts, the day after the Japanese attack the declaration of war passed with the sole opposition of a single representative.[136] Isolationists in Congress, finding themselves politically exposed, publicly acknowledged that their preferred policies would not have worked.[137] While Republican criticisms continued throughout the war, the focus was on inefficiencies in the war effort; they did not criticize the decision not to negotiate.[138] In public opinion, where even before Pearl Harbor a clear majority favored involvement in the European War, the Japanese attack basically ended all debate.[139] Moreover, wartime polling consistently recorded more than 50 percent of the American public opposed to even *discussing* peace with a government headed by the German army, with substantially higher majorities opposed to actually agreeing to peace on available terms or to negotiating with Hitler.[140]

The Missing Preventive Motivation

To the extent that the bargaining and war literature has developed an explanation for the Allied refusal to negotiate, it concerns commitment problems: the Allies fundamentally did not trust Hitler to abide by any war-ending agreement.[141] This concern, however, had nothing to do with the preventive motivation to fight. Although Germany had achieved great gains with respect to its opponents, those opponents did not expect Germany to become stronger with time. Without some belief that it was better to fight now than later, however, the preventive war mechanism is unable to account for the Allied refusal to negotiate.

The British leadership obviously recognized that victory over France made Germany far more powerful relative to Britain, even to the extent of creating a credible invasion threat. The problem with an argument grounded in shifting power, however, is that British leaders do not appear to have believed that capabilities would continue to shift against them. Hitler was not alone in believing that time favored Germany's opponents; British leaders agreed. Halifax, for example, argued that time was on Britain's side both before and after the declaration of war.[142] The collapse of France strengthened Germany substantially, but this accretion of power was a one-time thing rather than the start of a broader trend. In other words, when the War Cabinet debated policy after the fall of France, they were confronting a new strategic situation, but not necessarily a deteriorating one—no one raised the possibility that Hitler might be better able to invade the British Isles after a period of peace than he was in the months immediately after the French capitulation. Indeed, the belief that the German economy was overstretched, as well as the expectation that the United States would provide more effective assistance over time, provided reason to think that Britain would benefit from delay.

A similar problem exists with a preventive explanation for unconditional surrender. If Germany's rise was the primary concern, then it should be possible to reverse it by forcing the Germans back to something like their previous borders. After the defeat at Stalingrad, the Germans were no longer capable of conquering Europe militarily. While a deal with Hitler would have been unattractive, it could have saved the lives of millions on both sides who died in the final two years of fighting. It is of course a legitimate question whether Hitler would have been willing to accept a deal that returned Germany to its prewar frontiers— the answer almost certainly is no—but the fact that the Allies showed absolutely no interest in this sort of solution indicates that they were not primarily concerned with material commitment problems. Similarly, the constant harping on the nature of the Nazi regime, "Hitlerism," and Prussian militarism, which I discuss in greater detail below, is inconsistent with the basic commitment problem argument. If material power shifts were the primary concern, then we would not expect Churchill and Roosevelt to spend so much time focusing on the character of German

[131]

leaders. All these observations indicate that unconditional surrender cannot be explained by reference to a belief that German power was increasing over time.

Alliance Politics and Unconditional Surrender

Finally, a number of historians have highlighted the significance of alliance politics for the decision to demand unconditional surrender.[143] At Casablanca, Roosevelt explicitly raised the possibility that unconditional surrender would alleviate Stalin's fears of abandonment, which were likely to be heightened by the renewed Anglo-American decision to postpone the creation of a true second front in Europe.[144] The policy also had the benefit, at least from the American perspective, of postponing discussions of postwar spheres of influence, which in Roosevelt's view could only foster Allied disunity.[145]

The problem with this argument, however, is that it is more useful for explaining why the Allies decided to make the unconditional surrender policy public than why they decided on the more general policy in the first place. The Casablanca announcement was no doubt driven largely by alliance politics, but as the discussion of the historical origins of the policy above demonstrates, it was only a public statement of a decision that had been made some time earlier. It is certainly the case that alliance politics cannot explain the British refusal to negotiate when they had no allies, nor can it explain why American generals would advocate a fight to the finish even if Britain and the Soviet Union were knocked out of the war.

An interesting analogue to this point concerns criticism of the unconditional surrender policy, which was widespread both at the time and subsequently, but which focused almost entirely on its public nature rather than on the initial decision not to negotiate.[146] Contemporary criticism by American generals (including Eisenhower) and by Secretary of State Cordell Hull consistently was predicated on the belief that it hindered military victory by emboldening German resistance; the critics accepted the complete defeat of Germany and Japan as a goal, but believed that it would be easier to achieve if the Allies kept that demand secret for as long as possible.[147] Thus, for example, the propaganda experts who complained that announcing unconditional surrender would strengthen German resistance "were not necessarily opposed to the principle of total defeat—but they considered it a disastrous mistake for the president to announce it publicly."[148] Similarly, after expressing his opposition to unconditional surrender, Hull then clarified that he thought that "the most severe terms should be imposed" on Germany and Japan; he just wanted to reserve greater tactical flexibility with respect to Italy and the German satellites.[149]

[132]

Explaining Two Decisions Not to Negotiate

Ultimately, the unconditional surrender mechanism provides the only credible account of the Allied decision not to negotiate with Germany. I argued above that Germany's expansion was driven by a pervasive fear of decline, grounded both in Hitler's ideology and the window of opportunity created by German rearmament and the purge of the Red Army. The targets of this expansion, however, did not share Hitler's ideology, did not intend to do to Germany what Hitler feared, and thus concluded that Hitler was a madman who was out to conquer the world. Allied leaders thus explained German expansion in terms of a dispositional commitment problem: Germany by nature was expansionist and aggressive, and thus any strategy for peace that did not address German character was inherently flawed. This inference thus led directly to the conclusion that no peace short of unconditional surrender was acceptable.[150]

This first part of this chapter is sufficient to validate several predictions derived from the unconditional surrender mechanism. Consistent with hypothesis 2a, the demand for unconditional surrender was made by the targets of a preventive war. Moreover, none of the targets of this war shared Hitler's view of the world, and thus all were innocent in the sense meant by hypothesis 2b. Indeed, a number of obstacles prevented an accurate understanding of German policies. Hitler's views of the world were frankly bizarre, and despite the thorough depiction of them in *Mein Kampf*, contemporaries frequently understood them at best poorly.[151] The campaign to revise the Versailles Treaty further muddied these waters, as Hitler publicly backtracked from his true intentions. In this context, when Hitler claimed that German security required the acquisition of large amount of territory in Eastern Europe, contemporaries understandably saw these comments as at best a joke. Moreover, ideological differences complicated comprehension. The Western powers did not share his racial view of international politics, nor did they see Germany's territorial limitations as a threat to the survival of the German nation (which after all they saw as the primary threat to Europe); there is no indication that they ever seriously considered the idea that Hitler might think of his policies in anything approaching defensive terms. As for the Soviets, the communist belief in class conflict was fundamentally at odds with Hitler's racial theories. While the Soviets believed the capitalist powers to be doomed in the long run, there was nothing special about Germany in this regard, nor did Stalin expect the German people to support what Marxism-Leninism saw as a war for the interests of high capital.[152]

The remaining discussion thus focuses on several additional implications of the unconditional surrender mechanism. In particular, the theoretical discussion in chapter 1 predicts that the demand for uncondi-tional surrender will be justified in terms of the opponent's character and that the belief that the opponent is dispositionally aggressive will follow from the opponent's preventive war. In addition, the logic of this mecha-nism implies that leaders will see unconditional surrender as necessary only for the countries that actually initiated the preventive war, and not for allies of those countries that entered the war opportunistically or out of compulsion. Further, if the dispositional commitment problem argument is correct, then beliefs among leaders about the "source of the evil" (i.e., what parts of German government or society were responsible for the commitment to aggression) should influence openness to negotiation in possible contingencies such as the overthrow of Hitler by the army. On all of these points, the historical record is remarkably clear.

The Emergence of Dispositional Interpretations of German Aggression

Although Hitler's regime was viewed as distasteful and untrustworthy from the time he entered office, it was not seen as necessarily aggressive or as nearly the threat to Europe and the world that it was to become. In response to Hitler's behavior, however, attitudes hardened, with the re-sult that by the time of the Phoney War, and more intensely as the war continued, Allied leaders concluded that Hitler's Germany was disposi-tionally aggressive, and thus that only the complete restructuring of Ger-man politics and society would suffice to guarantee peace. Given the potential concern that public complaints about German character might be rhetoric for domestic political consumption, the examples I highlight to demonstrate these points come primarily from private discussions; many more examples, both public and private, are readily available.

Starting with British leaders, scholars often fail to recognize the degree to which key figures' views of Hitler and the Germans evolved over time because we focus on the period in which specific leaders had the greatest influence over policy. Thus, Chamberlain and Halifax—the architects of appeasement—are believed never to have recognized the menace of Ger-many, whereas Churchill—the stalwart lion of opposition—is believed always to have recognized Hitler as a particular danger to the world. Both views, in fact, are mistaken.[153]

Chamberlain and Halifax both sincerely hoped that limited conces-sions to Germany might prevent a second world war; the colossal failure of this policy has left them with the reputation of "decayed serving men" whose pusillanimity betrayed their country.[154] In both cases, however, Germany's continual aggression ultimately led them to conclude that no settlement was possible. While initially quite hopeful that appeasement would work, Halifax fundamentally revised his opinions of Hitler in response to Munich and to the subsequent invasion of rump Czecho-slovakia. During the Phoney War, he had the responsibility of respond-ing to those who wished to see a settlement; at the time he noted in his diary that

> to define your war aims precisely as people want would mean for me, if I spoke all my mind, that I wished to fight long enough to induce such a state of mind in the Germans that they would say they'd had enough of Hitler! And that point is not really met by talking about Cz[echoslovakia], Poland and all the rest of it. The real point is, I'm afraid, that I can trust no settlement unless and until H[itler] is discredited. When we shall achieve this nobody can say, but I don't think any "settlement" is worth much without![155]

During the critical War Cabinet meetings in May 1940, he repeatedly ex-pressed his doubts about Hitler's reliability; to the extent that he contem-plated peace, he thought of any possible deal as analogous to the Treaty of Amiens, which had allowed for one year of peace in the otherwise unbroken twenty-two-year war between England and Revolutionary and Napoleonic France.[156] In September he agreed fully with Churchill that a peace proposal by Sweden should be summarily rejected, despite the imminent threat of German invasion.[157] Chamberlain's views fol-lowed a similar trajectory. Prior to war, of course, he strove vigorously for peace and was willing to make significant concessions as the price to avoid war. One consequence of this approach, however, was that when his best efforts failed he had no doubt that it was because he had "come up against the insatiable and inhuman ambitions of a fanatic."[158] Espe-cially once the war began, Chamberlain and Churchill disagreed far less than is frequently believed.[159] In the War Cabinet discussions, he ac-knowledged that continuing discussions might mollify the French, but he believed that an actual approach to Mussolini "would serve no useful purpose," while "it was right to remember that the alternative to fighting on nevertheless involved a considerable gamble."[160]

As for Churchill, his persistent and prescient warnings about the Ger-man threat, even prior to Hitler's accession to office, belie the extent to which the rationale for these warnings changed over time. Initially, his argument was a traditional Realist one: Britain could not both allow Ger-many to rearm and expect it to abide by the humiliating terms of the Versailles Treaty, just as would be the case for any other country in Ger-many's situation. Thus, in November 1932, he argued in the House of Commons that "the removal of the just grievances of the vanquished ought to precede the disarmament of the victors. To bring about any-thing like equality of armaments . . . while those grievances remain unredressed, would be almost to appoint the day for another European war."[161] He returned to this point the following April—arguing that "as surely as Germany acquires full military equality with her neighbours while her own grievances are still unredressed . . . so surely should we see ourselves within a measureable distance of the renewal of general European war"—and with increasing frequency thereafter.[162] In general, rather than advocate unstinting opposition to German demands to revise Versailles, he favored early action, undertaken from a position of strength. As for Hitler, as late as October 1937 he noted in print that

> although no subsequent political action can condone wrong deeds or re-move the guilt of blood, history is replete with examples of men who have risen to power by employing stern, grim, wicked, and even frightful meth-ods, but who, nevertheless, when their life is revealed as a whole, have been regarded as great figures whose lives have enriched the story of man-kind. So it may be with Hitler.[163]

Moreover, once convinced that the Government had begun taking the steps necessary to rearm, Churchill noted that "in spite of the dangers which wait on prophecy I declare my belief that a major war is not im-minent, and I still believe there is a good chance of no major war taking place in our time."[164]

This attitude was not to endure, however. In his first speech after Chamberlain's triumphant return from Munich, Churchill warned against passivity, arguing that "this is only the beginning of the reckon-ing. This is only the first sip, the first foretaste of a bitter cup which will be proffered to us year by year unless by a supreme recovery of moral health and martial vigour, we arise again and take our stand for freedom as in olden time."[165] By the Phoney War, Churchill was of the opinion that Britain "must and should fight [the war] to a finish."[166] In a speech on October 1, 1939, he asserted that "how soon [victory] will be gained depends on how long Herr Hitler and his group of wicked men . . . can keep their grip upon the docile, unhappy German people. It was for Hit-ler to say when the war would begin; but it is not for him or for his suc-cessors to say when it will end."[167] As prime minister, he resolutely opposed negotiation, repeatedly returning to the theme of a German dis-position toward militarism.

Roosevelt's beliefs about Hitler and Germany also changed substan-tially in response to German aggression. Prior to Munich, he believed that it was possible and indeed desirable to co-opt Hitler and Mussolini to preserve peace in Europe.[168] Thus, for example, he was initially sup-portive of Chamberlain's appeasement strategy, and he was open to a proposal by Sumner Welles to organize a conference to seek a general negotiated solution to the world's problems.[169] For him, Munich was a decisive turning point; in particular, Hitler's aggressive style, captured most centrally in the rejection of a proposal that would have given him his entire demand with respect to the Sudetenland, provided evidence of extreme aggressiveness. During the Phoney War, Roosevelt was already referring to Hitler as a "nut" and a "wild man," and arguing that the war could end only with his death or removal, from either within or without.[170] At the same time, he started to worry about the possibility, highly improbable in retrospect, that Germany would launch a direct attack on the Americas, most likely moving from North Africa to Brazil.[171] Indeed, this concern persisted throughout the war, although Hitler had never expressed any interest in the Americas. There is thus ample evidence that Roosevelt sincerely believed that Hitler aimed to expand without limit. It is unsurprising to learn that at one point he confided to his closest adviser that he "shudder[ed] to think of what will happen to humanity, including ourselves, if this war ends in an inconclusive peace."[172]

Unconditional Surrender and Germany's European Allies

A further implication of my argument is that the demand for unconditional surrender should apply only or primarily to those countries that were waging a preventive war, and not to those countries' allies. In the context of Europe, this prediction implies that Italy and Germany's eastern satellites would be able to get better peace terms than Germany. While the official line—that all surrenders except for that of Finland were unconditional—is at odds with this prediction, a closer examination shows a clear difference between Germany (and Japan) and the Axis allies. In the United States, the early discussions of the Advisory Committee on Post-War Foreign Policy drew a distinction between Germany and Japan on the one hand and Italy on the other, leaving open the possibility of negotiation with the latter.[173] Likewise, the British and Soviets were both open to setting aside unconditional surrender with respect to the satellite states.[174] Among the Allies, President Roosevelt was the strongest opponent of any weakening of terms for Italy or the satellite states, but he justified his position not by arguing that these states were innately aggressive but on the grounds that he did not want to weaken the principle of unconditional surrender before it had been applied to Germany and Japan.[175] Moreover, when advocating unconditional surrender at Casablanca, he initially advocated applying it only to Germany and Japan; Churchill requested Italy's inclusion at the behest of the War Cabinet.[176] In the event, Italy's surrender—the first by any country on the Axis side—involved far greater flexibility than was to be shown to Germany.[177] Finland, as previously noted, was allowed to reach a negotiated settlement, while the remaining three German satellites—Hungary, Romania, and Bulgaria—were inevitably going to end up occupied and administered by the Soviets given the need to pass through their territories to reach Berlin. Overall, there is a clear sense that in the European war Germany constituted a unique evil that required a special and unusual response.

Acceptable Alternatives to Hitler?

A final microlevel prediction that follows from my argument concerns attitudes toward contingencies that never arose. Specifically, Allied leaders repeatedly confronted the possibility that Hitler might be overthrown and thus had to decide whether or not they would negotiate in such a scenario—this point was most salient in light of failed July 1944 military coup, but the steady stream of peace feelers, many promising to overthrow the existing government, meant that the issue was raised repeatedly over the course of the war. My argument would predict that the answer to this question would depend on what they believed to be the locus of evil in Germany. If, for example, the innately aggressive disposition was found only among Hitler and the Nazis, then it would be reasonable to negotiate with a military government after a coup. If the military or all of society was implicated, however, then replacing Hitler would not remove the obstacles to peace, and thus negotiation would be unacceptable. As Allied leaders differed in their diagnoses, this prediction thus implies that we should see variation in their openness to negotiation.[178]

History is indeed consistent with this prediction, at least in the case of the three Allied leaders for whom we have the most information. Roosevelt's views of the source of the evil expanded over time, so that by 1943 he was arguing that not just Hitler or the Nazis but the whole of German society was committed to aggression.[179] Thus, when rebuffing a proposed memorandum from the Joint Chiefs of Staff that would advance a clearer definition of unconditional surrender with an eye to weakening German resistance, he explained his stance by noting that a

> somewhat long study and personal experience in and out of Germany leads me to believe that German philosophy cannot be changed by decree, law or military order. The change in German philosophy must be evolutionary and may take two generations. To assume otherwise is to assume, of necessity, a period of quiet followed by a third world war.[180]

It is thus unsurprising that Roosevelt saw no room for negotiation with any possible German government.[181] Stalin similarly believed that there was something fundamentally wrong with German society, as he had expected the German proletariat to oppose the invasion of the Soviet Union; he thus doubted that even the multigenerational processes of reform that Roosevelt advocated would stamp out the German disposition to aggression.[182] Churchill, by contrast, believed that the root of the problem was Hitler, the Nazi Party, and Prussia; he tended not to blame the German military or the people more generally, for example in 1941 blaming previous British governments for not addressing German grievances and hence "goading them to Hitlerism."[183] As a result, while he was adamant that Prussia must be broken off from the rest of Germany after the war, and while he was committed to not negotiating with Hitler or any Nazi regime, he "thought that it would be going too far to say that we would not negotiate with a Germany controlled by the army."[184] This observable difference, which can be attributed to differing beliefs about the depth of the commitment to aggression in Germany, provides further evidence that beliefs about a German dispositional commitment to aggression really did drive the demand for unconditional surrender.

World War II was the worst war in history, spreading death, destruction, and poverty throughout Europe and beyond. An explanation for this destruction must account first and foremost for German aggression, and in particular the willingness to fight another major power war, so soon after World War I, to acquire vast tracts of territory in Eastern Europe. I argue in this chapter that the best explanation for this aggression is Hitler's fear of relative decline, grounded both in his ideology (which predicted that the territorially constrained German nation would experience fundamental decline relative to continent-sized powers, most importantly Russia) and in the window of opportunity created by German rearmament and Stalin's purge of the Soviet armed forces. This explanation provides an account of Hitler's expansive war aims, his repeated resort to risky diplomatic and military strategies (including most notably the willingness to invade the Soviet Union prior to the defeat of Britain), and the contrast between repeated attempts to work out a deal with Britain and the United States and the refusal to contemplate serious negotiations with the Soviet Union. Hitler's actions guaranteed that World War II would be extraordinarily destructive.

A full explanation, however, must also account for the Allies' refusal to negotiate with Germany. I argue that conventional explanations grounded most obviously in overoptimism or alliance politics provide an unconvincing account of Allied behavior. Instead, the unconditional surrender mechanism provides the best explanation. Hitler's preventive war was intended to save the German nation from annihilation. The Allies, however, did not intend to destroy it, and thus dismissed the

[139]

possibility that German behavior might be motivated by security concerns, instead concluding that Hitler, the Nazis, and in the extreme the German military and the whole of German society were dispositionally committed to aggression. Given this conclusion, no settlement that did not allow the Allies to fundamentally reform Germany would guarantee peace. It was this conclusion that meant that all peace proposals would be rejected, and that the war would end only with Germany conquered, divided, and occupied for multiple generations.

[5]

Additional Commitment Problem Cases

The Crimean, Pacific, and Iran-Iraq Wars

In-depth case studies of the Paraguayan War and of World War II in Europe both have provided support for my central explanations for the particularly deadly wars that I am most interested in explaining. This chapter supplements those case studies with minicases of the Crimean War, World War II in the Pacific, and the Iran-Iraq War. Although these cases are presented in far less detail than the previous two, they provide an additional opportunity to see both the preventive war and the unconditional surrender mechanisms in action. In all three wars, one participant had significant preventive motivations for fighting that contributed both to their decision to fight and to the way in which they conducted the war. Moreover, in the Pacific and Iran-Iraq Wars, the target of this preventive war refused to negotiate with its opponent, justifying its position in dispositional terms. The Crimean War by contrast provides a useful negative case, in that the preventive motivation behind British policy did not lead the Russians to refuse to negotiate, for reasons that turn out to be consistent with my theoretical argument.

The Crimean War

Although the diplomacy preceding the fighting was unusually complex, the Crimean War was a relatively straightforward case of a conflict driven largely by preventive motivations. The British observed Russia's seemingly inexorable expansion with trepidation, and for strategic reasons related to the Black Sea Straits found the logical next steps in that expansion extremely worrisome. When the Russians made an attempt to start to formalize their influence over the Ottoman Empire, the British saw a possibly final opportunity to stem the Russian tide before the Russians maneuvered themselves into Constantinople. They thus launched an aggressive war, with highly ambitious war aims that would have placed a clear limit on Russia's future ability to encroach on British interests. Although the British did not achieve their greatest aims, they did manage to force the Russians to step back from the kind of expansion that they might otherwise have achieved. This behavior is consistent with the preventive motivation for war.

In contrast to the other cases of preventive wars discussed in this book, the Russians did not respond to the British policy by demanding unconditional surrender. This development is consistent with hypothesis 2b, as the Russians clearly desired to do what the British feared and hence, although angered by British policy, did not explain British policy in dispositional terms and consequently remained open to a negotiated settlement.

A Quick Review of Events

As I discuss in further detail below, the central fact leading to war was the slow but sure decline of the Ottoman Empire, which raised the prospect of its eventual collapse.

The actual path to war was convoluted, proceeding from a French-initiated dispute over the Holy Lands through Russian counterdemands that, to outside eyes, seemed designed to render the Porte—the Ottoman court—a protectorate of Russia. Britain and France offered support to the Turks, who, emboldened, refused to back down even after the Russians without resistance occupied the Ottoman principalities of Moldavia and Wallachia. When the Russians refused to evacuate the principalities, the sultan declared war in October 1853, and a few months later the British and French followed suit.[1]

An Austrian ultimatum convinced the tsar to pull Russian troops out of the principalities, and as a result the Crimean Peninsula ultimately became the primary locus of fighting. After an extended siege, the French and British managed to capture the city of Sevastopol in late 1855; the Russians offset this defeat slightly with the capture of the Turkish fortress of Kars, east of the Black Sea. With Austria threatening to enter the war and with France quite ready to exit it, the two sides were finally able to agree on an armistice followed by a peace conference. The resulting conference consisted primarily of all other involved parties conspiring to force concessions on Britain, which by then constituted the primary obstacle to peace but was unable to continue the war without French support. In the final agreement, Russia returned Kars in exchange for Sevastopol, relinquished influence in the Ottoman Empire, and agreed to the neutralization of the Black Sea and to free transit on the Danube. These terms involved significant and painful Russian concessions that limited their direct influence over the Ottoman Empire, although they fell short of the more extreme war aims held by some in Britain.

Preventive Motives for British Policy

The basic problem in the Crimean War was not that the central participants were unsatisfied with the status quo, but that they feared with good reason that the status quo could not survive. In particular, both Britain and Russia (the actors on whom this discussion will focus) had reason to worry about the impending collapse of the Ottoman Empire. Russia was the dominant land power of Europe, having demonstrated in living memory the ability to project power all the way to Paris, but its navy, although not irrelevant, was weak. The Russians thus benefited from the existence of a large but weak and internally divided buffer to the south in the form of the Ottoman Empire; indeed, at the tail end of the successful 1828–29 Russo-Turkish War the Russians concluded that it was in their interest to limit their own territorial gains so as to preserve the Ottoman Empire, a policy that they retained up through the Crimean War.[2] From their perspective, the Black Sea Straits provided a useful defense against the naval power of Britain and France.

For the British, the straits also provided a useful defense. Britain had built a worldwide empire on the strength of its naval power; maintenance of that empire required that naval routes remain open. Russia was not an immediate threat to British naval preeminence—to take one example, Nicholas's renunciation of the favorable treaty of Hünkâr İskelesi, signed after Russia came to the Porte's aid in a crisis in 1833, constituted a realistic assessment that Russia gained nothing from having the option of sending its fleet into the Mediterranean.[3] This situation, however, merely acknowledged that geography had granted Russia limited opportunities for an active use of its fleet—the British feared that granted a warm-water port the Russians would gradually build up their navy to the point that it would threaten British dominance in the Mediterranean. From the British perspective, the Mediterranean was strategically significant not only because of the importance of the territories that directly bordered the sea but because it provided the most direct route for British connections to India. Indeed, significant Russian territorial gains in Anatolia, even in the absence of a significant fleet, would have threatened the overland route to India even had Russia not acquired a significant fleet—at a time when the Suez Canal had yet to be built. Moreover, by the time of the Crimean War British commerce had developed extensive interests

[143]

in the Black Sea region that would have been directly threatened should the Russians gain the ability to close off the straits.[4]

Thus from both sides' perspectives there were reasons to desire that the Ottoman Empire hold on, as a weak power unable to use its control over the straits to the detriment of anyone. The problem, however, was that the Ottoman Empire was obviously dying.[5] Internally decrepit and incapable of serious reform, the empire had suffered a serious blow to its prestige with Greek independence in 1830 and was threatened to the core immediately thereafter by an Egyptian rebellion, which returned as a serious threat a few years later. At the time of the war, then, few would have guessed that the Ottoman Empire was destined to survive another sixty-five years. At the same time, the strategic significance of Ottoman territories raised the specter of a serious war in the event of the Porte's collapse. In this context, it was unsurprising that all the Great Powers favored the preservation of the Ottoman Empire.

That said, Russia's desire for Ottoman survival was driven by balance of power concerns rather than sincere preferences: as I discuss in more detail in the next section, the Russians had both ideational and strategic reasons to want gains at Turkey's expense, with gains in Ottoman Europe, including Constantinople, the most salient. In this context, Tsar Nicholas's history of ruminating publicly on the consequences of Ottoman collapse—on the ostensible grounds that it was better to be prepared when the inevitable happened—tended to unsettle his neighbors. In early 1853, his repeated return to the issue in discussions with the British ambassador, Lord Seymour, raised the concern that Nicholas sought an agreement on partition as a prelude to actually carrying it out. The subsequent Menshikov Mission to the Porte, in which the Russians used threats of force to try to get the Turks to abrogate promises extorted from them by the French, further raised these concerns, especially as the Russians advanced views of their own capitulatory rights in the Ottoman Empire that were not supported by prior treaties.[6] An Austrian attempt, in consultation with Britain and France, to advance compromise terms faltered when Russian foreign minister Nesselrode advanced what came to be known as the "Violent Interpretation" of the Austrian proposal, which insisted on Russia's right to intervene in the Ottoman Empire to protect the Orthodox population, a right that others feared would render the Porte a direct protectorate of St. Petersburg.

This sequence of events thus convinced the British that Russia aimed at the destruction of the Ottoman Empire, from which it hoped to benefit by the acquisition of Constantinople and the straits. In the words of the Duke of Argyll, a member of the Aberdeen cabinet, "There was in the mind of all of us one unspoken but indelible opinion—that the absorption by Russia of Turkey in Europe, and the seating of the Russian emperor on the throne of Constantinople, would give to Russia an over-bearing weight in Europe, dangerous to all the other Powers and to the liberties of the world."[7] This logic is what we would expect were the preventive motive for war in action: the British feared that left unchecked the Russians would precipitate the collapse of the Ottoman Empire and benefit from the consequences, putting themselves in a position to threaten British interests around the world. War then was worthwhile if it could forestall such gains. The problem, however, was that under the status quo Russia already held an overbearing position with respect to the Ottoman Empire; if the tsar was committed to undermining the Otto-man Empire, the only way to prevent him from doing so would be to force Russia back.[8] British decision makers thus pursued war aims that would have involved a major reduction in Russian influence over the Ottoman Empire.[9] While the British ultimately had to abandon their most aggressive aims, they did so only once the departure of the French made it militarily impossible to impose the kind of defeat on Russia nec-essary to secure such gains, and even so they were able to force through terms such as the neutralization of the Black Sea and the limitation of Russia's ability to intervene in Ottoman domestic affairs that greatly re-stricted Russia's ability to impose unilateral changes. Indeed, the Crimean War marked the end point of Russian territorial aggrandize-ment into the Balkans; subsequently, Ottoman losses in Europe led to the emergence of new states whose loyalty to Russia was not guaran-teed. British behavior was thus clearly consistent with a preventive motivation.[10]

The Absence of Unconditional Surrender

According to the theory advanced here, demands for unconditional surrender are typically made in the context of preventive wars, when the target misunderstands the initiator's behavior. This section thus exam-ines that theoretical claim, as well as the broader linked question of why the Russians remained open to negotiation. Historians unsurprisingly do not generally spend much time evaluating nonevents, and thus there is no historical consensus (or even discussion) of why the Russians re-mained willing to consider negotiation. That said, consistent with my theoretical expectations, there is good evidence that the Russians sin-cerely desired to do precisely what the British feared. As a result, it is not surprising that Russian aims remained limited.[11]

That the Russians were open to negotiation throughout the war can be readily demonstrated, even if specific Russian war aims were not always clear.[12] Prior to the war, the Russians were consistently open to negotia-tions; for example, formally accepting the Vienna Note (albeit subject to

Nesselrode's "violent interpretation"). Once the war began, the Russians recognized that they would be forced on the defensive in the Black Sea but hoped to press an advantage in the Balkans; an Austrian ultimatum—made because of fears that a Russian advance would lead to uprisings of Slavs in Turkish territories that then would spill over into Austrian territory—compelled the Russians to abandon that aim. Rich, quoting Russian prince Gorchakov, summarizes Russia's war aims as the relatively limited desire "to reaffirm on a solid basis the religious immunities of our brothers of the Orthodox Church," which would require effective guarantees beyond just the highly debased word of the Porte.[13] At no point in wartime negotiations did the Russians indicate any refusal to negotiate, instead simply rejecting the terms that the British were willing to accept.[14]

The Russians had extensive experience with realpolitik, which aided them in understanding British policy. Perhaps the most critical figure here was Foreign Minister Karl Nesselrode, whose steady hand helped to limit the damage done when Tsar Nicholas entertained one of his speculative flights of fancy in which ideology trumped reality. Nesselrode's accurate prognostications even well prior to the start of the war, which were based on a deep understanding of international politics accrued over four decades of high-level service, provide a remarkable testament to his understanding of the working of international politics. Thus, for example, early in 1853 he predicted that if Russia went to war it would have no allies and would end up fighting a difficult war against both France and Britain (whose involvement at that date was far from obvious).[15] He was fully aware of the fears that Nicholas's speculative flights of fancy might provoke and worked assiduously behind the scenes to provide less unsettling interpretations of his monarch's more intemperate statements. As for the tsar, despite his tendency for verbal faux pas, he also fully understood that the collapse of Turkey would create many more problems than it would solve; he simply had come to believe that Turkey's collapse was imminent, whatever outsiders did.[16]

At the same time, however, the Russians sincerely desired Constantinople, which had strong historical, geographic, and cultural attractions. The tsars considered themselves to stand at the head of the "Third Rome," successor to the Roman and Byzantine empires; possession of Constantinople, capital or co-capital of both empires, thus held strong religious overtones.[17] Possession of Constantinople would also have provided substantial prestige to a country that saw itself very much as on the margins of Europe.[18] Moreover, the Russians saw themselves as the leaders of the Slavic peoples and thus took an active interest in the Christian populations of the Ottoman Empire; conquest of Constantinople would perforce bring about the end of Ottoman control over the Slavic populations of Europe. Finally, for reasons articulated above, the Russians had a strong strategic interest in acquiring Constantinople and the straits. Overall, then, the British fears of Russian intentions had merit.[19] In this context, Norman Rich asks "How justified were Turkish and Western fears of Russia?" and answers, "As the Turkish experience over the previous two centuries had shown—and as had the experience of almost every other neighbor of Russia—they were very justified indeed."[20]

In this context, it is unsurprising that Nicholas responded to the dispute with Turkey by formulating a plan that would have involved a quick naval strike to capture Constantinople and the Bosporus, with a force sent out shortly thereafter to secure the Dardanelles.[21] While military warnings about the improbable success of such an endeavor convinced him to abandon these plans for the eventual decision to occupy the Principalities, he retained the hope that the war might liberate Constantinople up until the combination of Anglo-French intervention and the Austrian ultimatum forced Russia entirely onto the defensive. Given this desire, rather than view British statements about the Russian threat to Constantinople as pretexts, the Russians took them at face value.

Overall, then, this case fits the commitment problem arguments well. The British had a clear preventive motivation for war, and consistent with that motivation they pursued large war aims that they were reluctant to abandon even in the face of military and diplomatic setbacks. As with the Argentines in the Paraguayan War, however, the Russians wanted to do exactly what the British feared and hence understood the British motivation for fighting. Given this understanding, they had no need to resort to a dispositional explanation for British policy and hence remained open to negotiation throughout the war.

World War II in the Pacific

The puzzle of the Pacific theater in World War II is why Japan started a war with the United States, an opponent that had an economy over five times Japan's size and a population nearly twice as large, especially when the Japanese were already stretched by an ongoing war in China and when they combined the attack on the United States with the invasion of British and Dutch colonial possessions in Southeast Asia. Although some have attributed this decision to simple irrationality, a more convincing explanation is that by 1941 the Japanese believed that peace with the United States was if anything riskier than war.[22] This logic fits perfectly with the preventive explanation for war, as does the audacity of the subsequent Japanese attack. At the same time, the American response to the Japanese attack on Pearl Harbor was to conclude that the Japanese must be insane. This inference led naturally to the conclusion that sustained peace with Japan would require the remaking of the Japanese political system; it was thus only when the Japanese surrender guaranteed that such a degree of reform would be possible that the war ended.

Japanese Expansion and the American Response

After the forced opening of Japan to the world in 1854, the Japanese launched an accelerated program of modernization that within a few decades would put them in a position to win wars against China in 1894–95 and against Russia in 1904–5. These victories established Japan as a great power with a sphere of influence on the Asian continent. In the 1930s, the Japanese began to expand that sphere through repeated incursions into Chinese territory in Manchuria, culminating in the 1937 invasion out of Manchuria into mainland China. That war ultimately turned into a quagmire for the Japanese, as they controlled large parts of mainland China but could not impose a final defeat on their Chinese opponents. Moreover, the Chinese adventure generated rifts with the rest of the world, and in particular with the United States, which by the second half of 1941 was the only great power not committed to the war in Europe. Cursed by dependence on the United States for resources, in particular oil, the Japanese eyed expansion into British and Dutch colonial possessions (most significantly present-day Indonesia, which had significant oil reserves), but feared that such a move would bring the United States into the war. An American oil embargo in the summer of 1941 changed the strategic calculus. Intense internal discussions that fall ultimately produced the decision that Japan would attack the United States absent an agreement that reopened the oil supply. When the United States demanded a complete withdrawal from China as a prerequisite for such a move, the Japanese decided that they had no choice but to invade the European colonial possessions. Given that decision, they decided also to attack the Philippines (an American possession) and the US fleet at Pearl Harbor in Hawaii.

The Japanese war plan called for a lightning strike south to acquire the colonies, after which they, conscious that they could not conquer the United States, would pursue a negotiated settlement on terms that would effectively exclude the United States and its allies from East Asia.[23] On the military side, their plans worked well: the Japanese rapidly overran Malaysia, Singapore, and the Philippines, carried out a damaging attack on Pearl Harbor, and were quickly moving into the Dutch East Indies. Diplomatically, however, the strategy failed, as the United States refused to negotiate, ultimately deciding to demand Japan's unconditional surrender.[24] While the Allies put primary importance on winning the war in Europe, they also prosecuted the Pacific campaign with increasing success and were poised by the summer of 1945 to launch an invasion of the Japanese mainland. The combination of the atomic bombs dropped on Hiroshima and Nagasaki and the Soviet entry into the war convinced the Japanese government that defeat was inevitable, and they thus surrendered, subject to the proviso, which the Americans accepted, that the emperor remain in at least a figurehead position. The war thus ended with the Allied occupation of Japan, which set the stage for the complete reform of the Japanese political system.

Japan's Preventive War

The puzzle of Japanese policy has been the ambition of its aims when contrasted to the relative pessimism that Japanese officials expressed about the likelihood of victory in a war with the United States.[25] Thus prior to the attack on Pearl Harbor and the Philippines, key military figures acknowledged that Japan lacked the capability to force the Americans to surrender, while Admiral Yamamoto advised Prime Minister Konoe that Japan would fare well in the initial six months to a year of war but that he had "no confidence, however, if the war continues for two or three years."[26] The navy—which would bear primary responsibility for fighting a war against the United States—remained skeptical throughout the critical discussions, with Admiral Nagano for example observing immediately after the imposition of the oil embargo that even in the event of an immediate attack "it was doubtful whether or not we would even win, to say nothing of a great victory as in the Russo-Japanese War."[27]

The problem, however, was that peace was no more attractive. Japanese scholars had long argued that the continual growth of American power into East Asia meant that Japan needed to expand simply to maintain the current level of relative capabilities.[28] By the late 1930s, the Japanese had a number of immediate reasons to believe that they were confronted with a narrow window of opportunity in which to act. Perhaps most significantly, the war in Europe effectively removed a range of probable opponents: both the British and the Soviets were fighting for their lives, while the French and the Dutch were occupied and hence ill-equipped to resist Japanese expansion into their colonies.[29] An end to the war in Europe would greatly reduce the Japanese opportunity to address its resource problems.[30] Moreover, those resource problems had worsened dramatically with the inability to bring an end to the war in China. The urgency of addressing the resource problems increased markedly when the United States imposed the oil embargo, as the Japanese were now confronted with a situation in which whatever slender chance of victory they enjoyed was vanishing rapidly.[31] At the same time, however, the American price for dropping the embargo—a complete withdrawal from China—stood to weaken Japan substantially, raising fears that the Americans would simply pocket the concession and then return to attack a few years later.[32]

At a number of points, Japanese leaders advanced arguments that could serve as textbook examples of the preventive motivation for war. In a critical discussion on October 30, 1941, the Japanese leadership was almost unanimous in its assessment that acceptance of American terms would eventually relegate Japan to the status of a third-rate power, economically dependent on the United States and its allies; doves were overwhelmed by the argument that "it [was] better to go to war now than later."[33] Similarly, a document prepared in consultation between the government and the military leadership prior to a meeting on September 6, 1941, argued that war was inevitable and that Japan was better off fighting soon:

It is historically inevitable that the conflict between [Japan and the United States] . . . will ultimately lead to war. It need not be repeated that unless the United States changes its policy toward Japan, our Empire is placed in a desperate situation, where it must resort to the ultimate step—namely, war—to defend itself and to assure its preservation. Even if we should make concessions to the United States by giving up part of our national policy for the sake of a temporary peace, the United States, its military position strengthened, is sure to demand more and more concessions on our part; and ultimately our Empire will have to lie prostrate at the feet of the United States.[34]

The same document advocated expansive war aims, specifically the elimination of all British, American, and Dutch influence in East Asia. Such a development would allow Japan to escape its resource predicament and would protect it even should American power continue to grow in the future. Indeed, the simultaneous attacks on the Malay Peninsula, the Philippines, and Pearl Harbor constituted an audacious undertaking that the Japanese recognized was quite risky. While the Japanese clearly contemplated a compromise peace, it was understood that this peace would be on terms that would involve American capitulation after serious setbacks in the war.[35] These ambitious aims, in contrast with the pessimism about Japan's military prospects, are clearly inconsistent with any explanation grounded in optimism, but they are what one would expect from a preventive war driven by fear of decline.

[150]

Unconditional Surrender

The Japanese strategy thus failed not because they overestimated their military chances but because their political gambit—hoping that the United States would lack the stomach to carry out the costly war to defeat Japan—did not pan out. Indeed, the American response was not only to refuse to negotiate on Japanese terms but to refuse to negotiate at all. Chapter 4 lays out the background to the decision to demand unconditional surrender, in which Allied discussions quickly concluded that negotiation with either Germany or Japan was unacceptable, although they waited some time to publicize that position and debated thereafter whether the public announcement was in fact the correct move.[36]

This case provides support for all of the hypotheses derived from the dispositional commitment problem. Consistent with hypothesis 2a, the US demand for unconditional surrender arose in response to Japan's preventive war. The justification for unconditional surrender was consistently advanced in dispositional terms. Thus, in his 1942 State of the Union address, Roosevelt observed "the militarists in Berlin and Tokyo started this war" and that the war would end only with "the end of militarism" in those countries; "there never has been—there never could be—successful compromise between good and evil." Elsewhere he promised that the United States was going to "strangle the Black Dragon of Japanese militarism forever."[37] The immediate reaction to Pearl Harbor, in Congress and among the public, was to conclude that the Japanese must be insane; this point was made most strongly by the isolationists who had previously opposed American entry into the war.[38] In contrast to Germany, where the obvious evils of the leadership seemed to provide some defense for ordinary German citizens, the American public saw all the Japanese people as fundamentally aggressive.[39] In this context, both the public and the leadership repeated the theme that there could be no negotiation with such an opponent throughout the war.

This argument is also borne out by the way in which the war ended. As the war turned against them, the Japanese, like the Germans, tried repeatedly to convince the Allies to reduce their war aims.[40] In the final months of the war, this strategy focused in particular on convincing the Soviets, who were still upholding an April 1941 neutrality pact, to serve as mediators, a strategy that was undermined by Stalin's interest in ensuring that the war lasted long enough to permit the Soviets to intervene and thereby gain a share of the postwar spoils.[41] The attempt to gain Allied concessions seemed to have some chance of success, given the extremely high expected costs of an invasion of the Japanese home islands. The Soviet entry into the war, which directly followed the atomic attacks on Hiroshima and Nagasaki, undercut this strategy, however, and forced the Japanese to capitulate. That said, the role of the emperor—who was philosophically the center of the Japanese political system—remained a potential sticking point. The initial Japanese acceptance contained the stipulation that the emperor would retain his political position in post-war Japan. Consistent with the dispositional argument, the Americans rejected this condition, despite the desire to avoid the horrendous losses of an invasion of Japan and to limit Soviet gains in East Asia, requiring instead that the emperor be subordinate to the Allied Supreme Commander.[42] It was only when Emperor Hirohito accepted this demand—abandoning his political role so as to at least save the imperial house—that the war actually ended.[43]

Overall, then, the Pacific War provides a further example of both commitment problem arguments in action. The Japanese launched an aggressive war, despite prevailing pessimism about their military prospects, because they believed that the decline that they would experience under peace would be if anything more risky. Consistent with the logic of preventive war, the Japanese aimed high and were willing to accept a startling amount of risk in pursuit of a peace that would inoculate them against decline, although they would have been open to negotiation on terms that prevented the decline from occurring. The Americans, however, inferred from the Japanese attack that they were confronted with madmen who could not be reasoned with. Given this conclusion, peace could come only through the reform of the Japanese political system, and the Americans consequently rejected any Japanese terms that preserved anything more than a figurehead role for the emperor. The war thus ended only with the surrender and military subjugation of Japan.

THE IRAN-IRAQ WAR

The Iran-Iraq War, fought between 1980 and 1988, is one of the longest and, at a per capita level, deadliest wars of the last two hundred years. What started nominally as a border war escalated into a conflict that proved remarkably difficult to end, with the Iranians in particular consistently rejecting settlement proposals. Eight years of fighting ultimately demonstrated that neither side had the ability to impose a decisive defeat on the other: although the Iranians managed to contain the initial attack and force the Iraqis back into their own territory, they were unable to achieve the sort of breakthrough that would allow them to dictate peace. Ultimately, then, the Iranians were forced to abandon their long-standing refusal to negotiate and accept a cease-fire.

This case presents a number of analytical difficulties. Given both the recency of the war and the autocratic nature of the participants, we do not have access to the sorts of details about each side's decisions that are available for most of the other case studies in this book. Most notably, information on the Iranian decision to negotiate, after years of rejecting talks, is largely limited to public statements in which the incentive for rhetoric is clear.[44] From a theoretical perspective, this case occupies a somewhat ambiguous (although for that reason potentially informative) role with respect to the dispositional commitment problem argument, as the Iranians clearly framed their eight-year refusal to negotiate in dispositional terms but ultimately backed off of their demands and accepted a negotiated settlement. That said, while the limits of available evidence mean that we cannot distinguish among competing explanations as definitively as we would like, the available evidence is certainly consistent with my argument.

A Brief History of the War

Iran and Iraq had long contested the exact location of their border, especially along the strategically significant stretch near the Persian Gulf. The 1975 Algiers Agreement between Saddam Hussein and the shah of Iran, at a time when Iraq was confronting a significant Kurdish uprising backed by Iran, appeared to settle this dispute in Iran's favor.[45] When the Iranian Revolution in 1979 deposed the shah, however, Saddam saw an opportunity to revise the settlement in Iraq's favor and thus launched an invasion. Although the Iraqis made significant initial gains, the Iranians managed to stop the Iraqi advance within the first few months, and by 1982 pushed the war back into Iraqi territory.

As Iraq's fortunes turned, Saddam began to signal a greater interest in peace. The Ayatollah Khomeini—the supreme ruler of the new Iranian government—categorically rejected negotiation, however, from the outset of the war averring that "we will not negotiate with them because they are corrupt. . . . Only if they surrender, for the sake of Moslems, we might consider something."[46] The international community initially viewed these statements as rhetoric, but the continued Iranian refusal to negotiate even after the Iraqis were expelled from Iran demonstrated Khomeini's sincerity. The war thus stretched on for eight years, during which the Iranians launched a series of offensives with the goal of ultimately imposing regime change on Iraq. The Iraqis, backed by the United States and by Middle Eastern powers anxious to limit the influence of revolutionary Iran, managed to hold off the Iranian invasions and achieved a significant strategic victory in 1988 by recovering the Fao Peninsula, the site of Iran's most significant victory. Shortly after this setback, the Iranians began to signal a greater willingness to negotiate; the war ended with Iran's acceptance

[153]

of a UN Security Council resolution that called for a return to the prewar status quo.

Saddam Leaps through the Window of Opportunity

Studies of this war are nearly unanimous in their agreement that Saddam launched a preventive war, although they also note that he overestimated the ease of Iraqi victory.[47] Iraq was substantially smaller than Iran both demographically and economically, and moreover was threatened by internal tensions related to both the recently rebellious Kurds and the majority Shi'a population. This disadvantage was central to the humiliating Algiers Agreement, and it could certainly be expected to be a feature of relations with Iran in the indefinite future. Meanwhile, intemperate Iranian commentary about spreading the revolution focused in particular on Iraq, given its proximity to Iran and its large Shi'a population, which could be expected to be more sympathetic to Shi'a Iran than the Sunni populations of the Gulf monarchies.[48] The Iranian Revolution certainly seems to have emboldened religious dissidents in Iraq.[49] These statements, combined with evidence of violent intentions such as the attempted assassination of Iraqi foreign minister Tariq Aziz (which may or may not have had direct Iranian backing), appear to have convinced the Iraqis in the first half of 1980 that war with Iran was inevitable.[50] At the same time, the failure of internal coups and of the American attempt to rescue the hostages in Iran demonstrated that Saddam could not count on anyone else to eliminate the Iranian regime for him.[51]

If war was inevitable, however, then there were strong reasons to prefer to fight sooner rather than later. Attempted counterrevolutionary coups forced the new government in Iran to purge both the military and the civil service, depriving the state of many of its most qualified leaders. Even before the purges, the revolutionary government had stripped the military of much of its advanced technology while overseeing widespread desertions.[52] The resulting chaos presented Iraq with an opportunity to achieve the sort of victory that would not be possible at any other time. At the same time, Iran was diplomatically isolated, having alienated the United States—its traditional superpower ally—through the taking of a large number of American hostages at the same time that it objected publicly to the Soviet intervention in Afghanistan.[53] Moreover, Saddam had managed to put down challenges from the Shi'a and Kurdish populations in Iraq, leaving Iraq well positioned to fight.

Debate exists about the exact extent of Iraqi war aims. Saddam's stated goals included Iraqi control over the Shatt al-Arab—the strategic waterway that had been in dispute with Iran prior to the 1975 agreement—and over several Persian Gulf islands as well as autonomy for several

[154]

minority groups in Iran, most notably the Arab minority in Khuzistan.[54] The territorial gains would have addressed a central strategic problem— Iraq had only thirty miles of coastline and thus was vulnerable to attacks on its oil exports. Moreover, several commentators have argued that the demand for autonomy for Iranian minority groups, which might appear as mere window-dressing, was a harbinger of plans to establish an effectively or juridically independent region including Khuzistan, which contained most of Iran's oil wealth, with the goal either of permanently weakening Iran or of precipitating the overthrow of the Khomeini regime.[55] Saddam explicitly disavowed any specific limits to Iraqi war aims, claiming that Iranian actions in border skirmishes prior to the war granted Iraq "additional rights," the extent of which he would only disclose at a later date.[56]

Unconditional Surrender Pursued and Abandoned

If the war initially arose primarily because of incentives to engage in preventive action, for much of its duration the primary obstacle to settlement was the Iranian refusal to negotiate. The most common explanation for this refusal is a domestic political one: fighting the war allowed the regime to consolidate the revolution, and they thus continued to fight until evidence of public unrest indicated that the war was no longer serving this purpose.[57] This unrest in turn is supposed to have followed from the high costs of the war and the military reversals that Iran suffered in 1987 and 1988. This argument is of course inconsistent with my theoretical arguments in multiple ways. For one, I argue that domestic politics cannot account for extended, high-intensity wars because society will catch on to the leaders' misbehavior and force a settlement. Moreover, I argue that this sort of nonnegotiation will follow from a sincere belief in a dispositional commitment problem. In this case, that argument would imply that the Iranians inferred from Iraq's preventive war that Saddam Hussein was by nature aggressive and hence that some degree of regime change was a prerequisite for settlement. Unfortunately, we do not have access to internal Iranian discussions of policy, meaning that a definitive resolution between these two explanations is impossible. That said, I highlight several reasons to believe that my argument provides a more convincing account of Iranian behavior than one grounded in domestic politics.

It is undeniably true that the war facilitated the consolidation of the revolution in Iran. The theoretical discussion in chapter 1 nonetheless raises two questions about a domestic-political explanation for Iranian behavior. First, did the Iranian leadership choose not to negotiate *because* they expected that war would help them to consolidate power, or did they prefer continued war for other reasons and gain a stronger hold on power as an ancillary benefit? Given that the regime would never admit this aim publicly, and given that we lack information about internal debates, we have no direct evidence that this motive drove the government's behavior. The existence of an alternative explanation for the decision to keep fighting thus could seriously undercut the domestic-political one. This point is especially true in light of the second question theory leads us to ask: if the regime was fighting for domestic political reasons, why was the Iranian public willing to put up with this behavior for so long?

The revolutionary regime certainly did not downplay the extent of suffering that was going on in the war with Iraq. Instead, it drew attention to its citizens' sacrifices, hailing the many dead as martyrs and even maintaining a "fountain of blood" (complete with dyed water) in Tehran to honor the fallen.[58] Indeed, Iranian tactics, which frequently relied on human wave attacks that used young men in the *Basij* as cannon fodder to offset Iraq's qualitative technological advantage, seemed almost designed to impose a high cost on society. Yet the majority Persian population remained broadly supportive of the war, with thousands volunteering for the *Basij*, even as military reverses meant that the date of final victory stretched ever further into the distance.[59] That they did so suggests that they shared the government's goals and were willing to pay the price necessary to achieve them. The continued popularity of the war, despite the leadership's openness about its high costs, provides a reason to believe that the regime was not continuing the war primarily for domestic-political reasons.

My argument by contrast would claim that the Iranian refusal to negotiate followed from a sincere belief that Saddam was dispositionally aggressive, which in turn was motivated by Iraq's preventive war. Again a definitive conclusion about the validity of this interpretation is impossible given the dearth of evidence, but there are a number of suggestive points. The Iranian leadership certainly justified the war in dispositional terms, repeatedly referring to Saddam Hussein and the Ba'ath Party as "corrupt" and un-Islamic.[60] Moreover, its demands were consistent with the dispositional commitment problem. From the opening days of the war, the Iranians specified that a minimal condition for peace would be that Saddam Hussein be tried as a war criminal, for which his removal from power would be an obvious prerequisite. As the war continued, the Iranians demanded the purging of the entire Ba'ath Party from the Iraqi government, a requirement consistent with the belief that the Ba'athists were dispositionally aggressive—indeed, Foreign Minister Velayati repeatedly compared the Iranian efforts to the Allied policy of unconditional surrender during World War II.[61] More generally, they clearly

[156]

preferred to try to spread the revolution to Iraq by not only removing the offending regime but replacing it with a kindred Islamic government. Moreover, after accepting the cease-fire, Khomeini repeatedly warned the Iranian people that the matter was not closed and that the Iranian people should be prepared for another attack, consistent with the belief in a dispositional commitment problem.[62]

As for the origin of the belief that Saddam Hussein's Iraq was by nature aggressive, the preventive nature of Iraq's attack is obviously consistent with my argument. My theory contends further that this sort of inference is particularly likely when the target of the preventive war does not intend to do what the initiator fears. Saddam's most central fear was that once consolidated in power the Iranians would seek to actively spread the revolution through an invasion of Iraq. This fear was not unreasonable given the stated Iranian desire to see their revolution spread throughout the world and given the support and encouragement that the revolutionary regime provided to disaffected groups in Iraq. Thus, for example, on the day that the revolution triumphed, the Ayatollah Khomeini proclaimed, "We will export our revolution to the four corners of the world because our revolution is Islamic; and the struggle will continue until the cry of 'There's no god but Allah, and Muhammed is the messenger of Allah' prevails throughout the world."[63] That said, the Iranian regime repeatedly consistently disavowed any intention to spread the revolution through force, instead claiming that they merely wished to see others achieve an Islamic revolution in their own countries.[64] Skeptics of course will observe that the Iranians likely would have made this claim whatever their true intentions, although it is true that they did not take advantage of subsequent opportunities to attack Iraq, for example during the period immediately after the Persian Gulf War when Saddam's hold on power was threatened. Absent better information about their ultimate plans, therefore, it is impossible to prove definitively that they were innocent in the sense meant by hypothesis 2b. It is not implausible that they did not intend to do what Saddam feared, however, and my argument predicts that were more information on Iranian intentions to emerge, it would turn out that they did not have the intentions that Saddam ascribed to them.[65]

The obvious puzzle for this perspective is why the Iranian government abandoned its refusal to negotiate and accepted a resolution that left Saddam Hussein and the Ba'athist Party in power. Here again the answer must be speculative, but the best available explanation is that the Iranians came to realize that they simply lacked the capacity to impose the defeat that was necessary to force Saddam from power. After five years of seeking to break through Iraqi defenses, the loss of their most significant military gain—the Fao Peninsula—and the evidence of renewed Iraqi strength in 1988 indicated that Iran would never break through and achieve the military victory it sought.[66] To the extent that we have information about the internal discussion prior to the cease-fire, it is consistent with this interpretation: the Ayatollah Khomeini described accepting the peace as "drinking a chalice of poison," which he did only because victory was impossible without a major military buildup and the development of "laser and nuclear weapons" of the sort that were entirely out of Iran's reach.[67]

Overall, then, although the limits of available evidence mean that conclusions cannot be as definitive as they are in other cases, the Iran-Iraq War is certainly consistent with my argument. Presented with a window of opportunity, Saddam Hussein launched a preventive war that held out the potential to lock in Iraq's historically unprecedented relative advantage over Iran. When this attack miscarried, the Iraqis found themselves fighting a defensive war against an adversary that refused to contemplate settlement with the existing Iraqi government, claiming that it was corrupt and hence not an acceptable negotiating partner. While the Iranians ultimately abandoned their aims and accepted a compromise peace, they did so only after five years of enormously destructive warfare when it became apparent that they could never impose the regime change that they sought.

[6]

Short Wars of Optimism

PERSIAN GULF AND ANGLO-IRANIAN

Thus far, I have examined only long, high-intensity wars. The repeated appearance of commitment problem concerns among the central motivations to fight in these cases provides strong evidence in favor of the claim that concerns about an adversary's inability to commit—whether because of adverse shifts in relative capabilities or because of a belief in the opponent's dispositional commitment to war—produce unusually destructive wars. As I noted at the outset, however, a research strategy that involves examining only large wars is problematic. Most fundamentally, if similar commitment concerns are also present in more minor wars, then it is hard to argue that they account for the difference between bigger and smaller conflicts. Moreover, an explanation for large wars becomes more compelling to the extent that we also have a coherent explanation for more limited conflicts, such as those provided by the informational and principal-agent mechanisms.

This chapter and the next thus examine a set of wars that were more limited in severity and, typically, duration than the wars discussed thus far. In every case, either the informational or the principal-agent mechanism turns out to have been most significant in bringing about the war, while significant commitment problem concerns were absent. This chapter examines the Persian Gulf War of 1991 and the Anglo-Iranian War of 1856–57, each of which I argue was driven by the informational mechanism.

THE PERSIAN GULF WAR

The optimism about international peace that followed the end of the Cold War did not last long—the Iraqi invasion of Kuwait in the summer of 1990 and the subsequent war between Iraq and a coalition led by the United States demonstrated that the use of force would remain a feature of the new international system. Whereas in the Iran-Iraq War, Saddam Hussein launched his invasion in response to a fear of relative decline, in this case closer analysis reveals that his attack was a consequence of overoptimism about the likelihood that outside powers would intervene to reverse his fait accompli and the degree to which Iraq could success-fully defend its conquest in the event of war. It is thus not surprising that the war followed the logic of the informational mechanism quite closely. Starting from quite disparate expectations about the likelihood and con-sequences of war, the participants advanced incompatible demands that precluded a peaceful settlement. Indeed, this process occurred twice, first with the failure of negotiations between Iraq and Kuwait prior to the Iraqi invasion, and then with the inability to reach a negotiated settle-ment to the resulting crisis between the Iraq and the coalition that mobi-lized against it. Ultimately, it took war to bring these expectations into alignment: once the fighting started, and especially once the shift to ground combat demonstrated that the Iraqi army was not capable of re-sisting the opposing coalition effectively, Saddam Hussein rapidly scaled down his demands, ultimately accepting a settlement far worse than the one on offer prior to the outbreak of violence.

A Brief Review of Events

Iraq emerged in 1988 from its eight-year war with Iran militarily and economically exhausted. In the course of that conflict, the country had accrued significant debts, including loans from many of its neighbors that it subsequently resisted repaying, on the rationale that its armed opposi-tion to Iran had served the general interest of the Arab Gulf states. The refusal of these governments to forgive the loans after the war thus was a source of significant discord, as was overproduction of oil by some of the smaller OPEC countries—notably including Kuwait—which drove down the price of Iraq's primary export and thus hindered the country's efforts to recover financially and rebuild after the war.[1] At the same time, the Iraqis had long claimed, on the basis of the internal organization of the Ottoman Empire, that Kuwait was legitimately part of the province of Basra and hence should be governed by Iraq.[2] When the Kuwaitis refused to address Iraqi grievances, war thus was an attractive option, and the Iraqi army duly invaded on August 2, 1990, completing the conquest of Kuwait within a day.

The first Bush administration in the United States reacted strongly to the Iraqi invasion, helping to secure a series of UN resolutions denouncing Iraq's actions and organizing a coalition of thirty-four countries who sent soldiers, first to defend Saudi Arabia against potential further Iraqi expansion and then to compel an Iraqi withdrawal from Kuwait. When the imposition of economic sanctions failed to induce compliance, the coalition moved first to aerial bombing and then eventually to a ground war. Impartial observers had predicted high potential death tolls on both sides, but American technological superiority, a superior coalition strategy, and the reluctance of most Iraqi soldiers to actually fight resulted in a far more one-sided conflict than many had expected. Rather than launching an attack directly into Kuwait, the coalition instead flanked the Iraqi army by invading from Saudi Arabia into southern Iraq, from where it was in a position to cut off communication between Baghdad and Iraqi forces in Kuwait. The success of this attack forced the Iraqis into a rapid withdrawal at high cost, with the coalition demolishing units retreating along the highway from Kuwait City to Basra. The course of fighting forced Saddam into a rapid political retreat and even opened up the possibility of a march on Baghdad, which the Bush administration however opted not to pursue. The two sides thus reached a political settlement in which Iraq not only recognized Kuwait as sovereign and independent but made a slew of additional concessions, including the imposition of a UN monitor force, the destruction of its weapons of mass destruction program, and the payment of reparations to Kuwait.

A number of questions follow from this brief history. Why did Iraq take the gamble of invading Kuwait? Why did Saddam refuse to back down even after a broad coalition had formed to compel withdrawal? And, most important from the perspective of this project, why, given the sharply contrasting positions that each side established prior to war, did fighting end so quickly? To preview findings, the war conformed well to the predictions of the informational mechanism. Although the Americans arguably underestimated the ease with which they would win, they still expected to do far better than the Iraqis believed they would. Meanwhile, Saddam Hussein had serious (and not unfounded) doubts about the willingness of the United States and others in the West to actually fight, the strength of American resolve once war began, and the ability of the coalition to expel Iraqi forces from Kuwait at anything approaching acceptable costs. Given this divergence in expectations, prewar negotiation failed to identify a mutually acceptable political settlement. Once fighting began, both sides learned about the other's capabilities and resolve, and the new information forced Saddam to acquiesce rapidly to demands that he had ruled out as unacceptable prior to the war.

The Sources of Optimism

The central disagreement for the Persian Gulf War was between the United States and Iraq, but it obviously followed a prior disagreement between Iraq and Kuwait. A few words on Kuwaiti policy will thus be useful, as there is reason to believe that the entire conflict might not have happened had the Kuwaiti leadership understood that Saddam was willing to invade. The two countries had a number of disagreements—including the aforementioned loan disputes, Iraqi sovereignty claims, the possibility of Iraqi usage of the Kuwaiti islands of Warba and Bubiya, Iraqi grievances about Kuwaiti overproduction of oil, and disagreements related to the Rumaila oil field—on which the Kuwaitis consistently refused to make any significant concessions.[3] This unyielding negotiating position was in large part a function of confidence, grounded in reports from the United States and from Arab intermediaries—most notably Egypt's Hosni Mubarak—that Saddam was bluffing and would not in fact invade.[4] From the opposite perspective, however, this situation meant that an Iraqi government that was willing to resort to force, and that enjoyed overwhelming military superiority, was confronted with an opponent whose demands were unreasonable given the likely outcome of a war. The ensuing fighting amply demonstrated that Saddam was willing and able to enforce his demands, but his capability and resolve could only be credibly demonstrated to the Kuwaitis by the actual use of force.

Of course, invading Kuwait was attractive only so long as the Iraqis had reason to think that they could get away with doing so. The available evidence, however, indicates that Saddam did in fact believe that the consequences of an invasion for Iraq would be mild. This expectation was grounded in three separate beliefs: that the United States lacked the resolve to fight over Kuwait; that even if the Americans were willing to fight the Saudis would not be willing to supply the bases that would be needed for an effective response; and that in the unlikely event of a war the Iraqi army would be able to impose significant casualties on its opponents, which in turn would lead the Americans (and hence the broader coalition) to back down. It is thus worth spending some time discussing the origins of these beliefs.[5]

Expectations of American irresolve—whether with respect to the willingness to oppose the Iraqi conquest or the willingness to stick it out in a war once fighting began—stemmed in large part from a belief that the American people, and thus the government that represented them, were unwilling to suffer significant casualties, especially in pursuit of political aims in peripheral parts of the world.[6] Thus, for example, in his first and only meeting with US ambassador April Glaspie, on July 25, Saddam observed that "yours is a society which cannot accept 10,000 dead in one battle."[7] This belief was not new. A decade earlier, in the face of fears of Iran engendered by that country's revolution in 1979, Saddam concluded that relying on the United States for protection was a losing proposition, as the Americans had found that providing such protection was too costly and risky a venture to undertake.[8] Likewise, at a meeting of the Arab Cooperation Council in February 1990, he had observed that "all strong men have their Achilles' heel. . . . We saw that the U.S. as a superpower departed Lebanon immediately when some Marines were killed."[9] Moreover, Saddam increasingly came to think that his opponents cared more about his citizens than he did, and that images of Iraqi suffering would undermine public support for the war even in the event that his armies were unable to impose significant casualties on coalition forces. Thus, for example, at one point during the crisis the Iraqi leader predicted that CNN coverage of the first two bombing runs on Baghdad would generate enough public pressure on the administration that Bush would have to call off the air war.[10]

The belief that the Americans lacked the resolve to respond to the invasion with force gained support from the policy of "constructive engagement," under which the United States tried to remain conciliatory in the face of Iraqi provocations in the expectation that moderation would turn Iraq into a regional partner, if an admittedly sometimes difficult one.[11] In this regard, the criticism that Ambassador Glaspie endured for not providing a stronger deterrent threat—by some accounts giving Saddam a "green light" to invade—is overblown, as she was simply carrying out the administration's policy of trying to avoid directly antagonizing Saddam.[12] For advocates of constructive engagement, the policy held out the prospect of turning the regime with the largest army in the Middle East into a responsible member of the international community. This policy had the downside, however, that the United States made no clear deterrent threats, thus giving the Iraqis reason to believe that, although not welcomed, their invasion would not be forcefully opposed.[13]

Even in the event of a ground war, Saddam had reason to think that Iraq might emerge, if far from unscathed, politically victorious. The Iraqi army had gained an incredible amount of experience in the eight-year war with Iran, and the elite Republican Guard units were expected to pose a significant challenge to even well-trained Western units, while the huge size and impressive armament of the Iraqi general army—the fourth-largest in the world, larger than that of the United States—meant that even absent elite training it would pose a formidable obstacle.[14] Meanwhile, by the time the air war broke out, the Iraqis had had almost six months to reinforce their defensive line in Kuwait with minefields,

[163]

oil-filled fire trenches, sand embankments, and other obstacles.[15] In this context, Saddam could quite reasonably believe that any attempt to push the Iraqis out of Kuwait would prove sufficiently costly for the coalition that public opinion would force a premature end to the campaign.

A less-recognized basis for Iraqi overoptimism was the belief that the Saudis would not invite American soldiers to operate out of their territory. Thus, for example, Saddam reportedly ordered the occupation of all of Kuwait, rather than just disputed border areas, in the expectation that doing so would preempt a Kuwaiti invitation to the United States, which in turn would leave the Americans with no place to establish bases.[16] It was in keeping with this expectation that the Iraqis indicated a desire for an Arab solution to the problem, which in practice presumably would have involved significant concessions to Iraq, likely financed by the Saudis and others in the Gulf.[17] In practice, this expectation was undone by standard security dilemma fears: the Iraqis have consistently professed not to have had any plans to attack Saudi Arabia, but even if their intentions were limited the Saudis could not know that for certain, while the massive Iraqi army in Kuwait obviously was capable of continuing its offensive southward. It was this fear that induced King Fahd to accept the significant domestic political costs associated with inviting the United States and other Western powers into Saudi Arabia.[18]

The Evolution of Beliefs

Political disagreements and a difficult financial situation provided Iraq with reason to consider an invasion of Kuwait, and the weakness of the Kuwaiti defenses, the refusal of Kuwaiti leaders to make significant concessions, and a belief in American irresolution and inability to act indicated that the consequences of an invasion would likely be minimal. In this sense, Saddam Hussein had good reason to be quite confident when he ordered the invasion of Kuwait. Subsequent events, of course, would belie that initial optimism. Some bases for optimism had to be abandoned quite quickly: the Saudis took only five days to decide to invite in American forces. Others, however, persisted.

In particular, the belief that the United States was fundamentally irresolute continued to provide a reason for optimism, even as world opinion turned quickly against Iraq. The confidence in irresolution can be seen in Saddam's prewar diplomatic strategy, which consisted of repeating with great frequency the claim that American resolve would not withstand the first few encounters.[19] These beliefs can only have been strengthened by the internal divisions in the United States, where only a slender majority of the public supported the threat to use force and many in Congress were arguing vociferously for greater time to allow sanctions to take effect and calling on Congress, in the words of Senator Edward Kennedy, to "stop this senseless march to war."[20] When former secretary of state Cyrus Vance testified that in the event of war the United States would find itself "virtually alone in a bitter and bloody war that will not be won quickly or without heavy casualties," he was capturing—and broadcasting for anyone to hear—the sentiments of a large portion of the American public that believed that war would be a disaster.[21] President Bush was willing to go to war even without the support of Congress, but that kind of willingness was impossible to demonstrate credibly before the fact, while the Iraqis could listen to Democratic senators arguing that even a few thousand casualties in the liberation of Kuwait would be too high.[22]

The American leadership was fully aware that Iraq might not take American threats seriously, given the incentives to bluff.[23] Indeed, at the final meeting between leaders on the two sides, Secretary of State James Baker started his comments by observing to Iraqi Foreign Minister Tariq Aziz, "You think we are bluffing."[24] This task was complicated, however, by the need to maintain the domestic and international coalition for war, which involved demonstrating that all possible efforts had been made to reach a diplomatic settlement. Thus, regional experts reported that Bush's proposals for last-minute talks between Baker and Aziz—undertaken to demonstrate that all efforts to reach a peaceful solution had been taken—would be interpreted in Baghdad as evidence of irresolution, no matter how frequently Bush insisted that the American position was nonnegotiable.[25]

At the same time, Bush administration officials were growing more confident, with Bush in particular coming to believe that the United States could "knock Saddam Hussein out early."[26] A critical development was the discovery that, contrary to initial beliefs, the ground on the Saudi-Iraqi border west of Kuwait was firm enough to support tanks and large transport vehicles, thus permitting a ground war plan based on maneuver and envelopment rather than one based on a costly frontal assault on prepared defenses in Kuwait. As a result, whereas in October military and political leaders were reporting that "if war comes, its human, economic, and political costs are likely to be high" and that "there is little prospect of winning a neat 'victory' in such a conflict," by January Bush was operating on the basis of military estimates that put the likely toll below two thousand casualties.[27] Meanwhile, Bush was increasingly convinced that the Iraqi government's hold on power was tenuous, at one point in his diary comparing Saddam Hussein to Nicolae Ceauşescu, the Romanian communist dictator whose decades-long rule disintegrated in 1989 in the face of a general uprising.[28] Overall, then, American decision makers believed that war might provide significant benefits beyond the liberation of Kuwait. In this context, as the crisis developed American demands for unconditional withdrawal rose to include other demands such as reparation payments to Kuwait and the weakening or elimination of Iraqi WMD programs.[29]

Beliefs and Bargaining Positions Once Fighting Began

Ultimately, large shifts in bargaining positions only occurred once the war began, and especially as the Iraqis came to realize that their expectations, and hence their strategies, were flawed. Before the war started, Saddam repeatedly averred that Iraq would withdraw from Kuwait only as part of a general Middle East peace agreement, and that absent that effectively impossible condition Kuwait was now an inviolable part of Iraq.[30] He held to that position even as the deadline for withdrawal approached, apparently believing that American irresolution would manifest itself either in a decision not to attack or, after initial clashes, in a reluctance to pay the costs necessary to evict the Iraqi army from Kuwait. Once the bombing began, he continued with the same basic strategy of trying to split the coalition by linking its activities to Israel—now by launching missile attacks on Israel designed to provoke an Israeli response—while promising to impose unacceptable casualties on coalition forces should an invasion take place.

Diplomacy did not stop once the bombing began, although it took on a new character. Particularly illuminating were repeated Soviet peace efforts in the period immediately prior to and during the war, which induced the two sides to reveal their bargaining positions.[31] These peace efforts foundered initially on Saddam Hussein's refusal to withdraw from Kuwait on anything approaching the terms contained in prior Security Council resolutions. Even when after almost a month of bombing the Iraqis indicated a willingness to accept Security Council Resolution 660, which called for unconditional Iraqi withdrawal, their acceptance was subject to extensive qualifications that guaranteed American rejection. This position changed with the approach of the ground war, however. In the days prior to the February 24 start to the ground assault, the Iraqi government made an increasing range of concessions, ultimately agreeing to a Soviet cease-fire proposal in which the Iraqis would withdraw unconditionally, removing all forces from the country within three weeks.[32] By this point, however, the success of the air war and the failure of Iraqi attempts to split the coalition increased confidence in the Bush administration, which now concluded that the Iraqi army was caught in a trap. In response, Bush raised American war aims to include not only the liberation and restoration of Kuwait but the substantial weakening of the Iraqi army. In line with this goal, he stated that peace was on offer only if the Iraqis would agree to complete withdrawal within one week, which would have required the abandonment of almost all of the army's heavy equipment. As a result, substantial Iraqi concessions—to terms that would have guaranteed peace prior to the start of fighting—were no longer sufficient to purchase a cease-fire. At the same time, however, the vast disparities between the two sides' demands prior to fighting had narrowed substantially.

Peace ultimately came only after the four days of ground war brought about the weakening of the Iraqi army that the Americans sought. The flanking of the Iraqi army in Kuwait precipitated that force's headlong withdrawal into Iraq, during which the coalition air forces were able to annihilate any retreating heavy forces, although key Republican Guard units, which never had been deployed to the front lines, were able to escape back toward Baghdad. At this point, Bush declared victory, announcing a unilateral cease-fire that the Iraqis were only too happy to accept. In practice, the specific terms of that cease-fire were worked out in a meeting between opposing generals at Safwan on March 3, after which the Iraqis begrudgingly acquiesced to additional Security Council Resolutions that codified the political terms of Iraq's defeat, including the establishment of no-fly zones over southern and northern Iraq in which the Iraqi air force was not permitted to operate and the creation of an intrusive monitoring regime to identify and eliminate Iraq's chemical, biological, and nuclear weapons programs.[33]

A separate but important question is why, given the success of the ground war, the Bush administration ruled out an escalation of the war to ensure regime change in Baghdad, a possibility that was discussed both at the time and subsequently. The decision to stop the war with Saddam still in power arose primarily from two concerns: that the coalition would collapse if the United States were to attempt to move beyond the mandate provided by the Security Council, and that an attempt to overthrow Saddam would involve the American forces in an indecisive hunt through hostile regions of Iraq that would expose them to urban guerrilla warfare that would inflict the high level of casualties that a more restrained military strategy had managed to avoid.[34] Thus, having attained their primary political aim and having furthermore substantially weakened Iraq, the Bush administration saw no further goals that could be achieved at acceptable cost.

After six months of inflexibility, then, both sides' bargaining positions shifted quite dramatically once the bombing started, and especially once the fighting intensified with the move toward the ground war. These developments are entirely consistent with the predictions of the informational mechanism: while credible signaling was difficult prior to the outbreak of costly fighting, the course of events once fighting began demonstrated that Iraqi expectations were profoundly mistaken and thus justified a dramatic shift in political demands. This process occurred quickly: even before the ground war rendered a continued Iraqi occupation of Kuwait impossible, Saddam had indicated his willingness to concede the central issue at stake in the war. Consistent with theoretical expectations, in a war in which divergent expectations led both sides to anticipate a relatively quick victory, only a relatively brief period of fighting was necessary for the two to identify a settlement that they preferred to continued war. Moreover, demands shifted more rapidly as the war intensified, with the shift to the ground war rapidly bringing about near-total Iraqi capitulation despite the survival of the elite Republican Guard units. Overall, then, we have a case in which divergent expectations about the likely course and outcome of conflict led to a war, but once that war began both expectations and bargaining positions shifted rapidly in a manner that allowed for quick settlement.

The Secondary Significance of Alternate Mechanisms

In contrast to the strong evidence of the relevance of the informational mechanism, the commitment and principal-agent mechanisms were of at best secondary significance. One possible commitment problem argument would be that the incorporation of Kuwait would increase Saddam's capabilities to such an extent that he would be able to engage in further aggression at a later date. While Bush administration figures certainly worried that Iraq might engage in further aggression, in particular with a further strike against Saudi Arabia, they did not express concern that control over Kuwaiti oil would facilitate such aggression. Indeed, the most worrying scenario was an attack on Saudi Arabia immediately after the initial invasion (before coalition forces had appeared in the region), before Iraq would have had an opportunity to turn Kuwaiti oil into a strategic resource. Instead, intervention was undertaken primarily because of the economic cost of allowing the consolidation of oil production under hostile governments and because of a sincere belief that permitting aggression to succeed would undermine international stability.

A separate argument about shifting power concerns the American insistence that Iraq abandon its heavy armor in Kuwait as a condition for peace. This requirement clearly was designed to weaken Iraq, and Saddam's reluctance to agree even after he had conceded the central political issue—Kuwait's independence—indicated that it was precisely this anticipated shift that prevented a settlement prior to the ground war. In chapter 1, however, I noted a distinction between beliefs in long-term decline that are a feature of preventive wars and these sorts of intrawar military shifts, as with the anticipated weakening of

[168]

the German Wehrmacht after the encircling of the Sixth Army at Stalingrad. The anticipated destruction of an entrapped force certainly may hinder settlement, but it will not produce the dynamics that make preventive wars difficult to resolve over an extended period of time, largely because the anticipated shift is likely to happen quickly. Thus in this case it took only four additional days of fighting for the Iraqi army to suffer the losses that it would have incurred through acceptance of American terms, after which this problem ceased to exist as an obstacle to settlement. It is for this reason that I argue that these sorts of intrawar shifts generated by anticipated military defeat do not invoke the broader dynamics of preventive wars.

A principal-agent interpretation is even weaker. As an absolute autocrat, Saddam worried about the possibility that he might be violently overthrown, either by his own army or by a Shi'a uprising, likely with Iranian backing. Indeed, he took extensive precautions against such a possibility throughout his reign—the steps that he took to forestall overthrow by the military, including executing prominent generals and limiting the military's training in urban warfare, rendered his regime almost totally coup-proof, while a repressive internal system prevented more general uprisings.[35] He thus had no reason to use a war to cement his hold on power; indeed, defeat in war was one of the few developments that might significantly have threatened his hold on power, as the Shi'a and Kurdish uprisings after the defeat in Kuwait demonstrated.

Likewise, it is hard to sustain a diversionary argument about the coalition response. While disagreement about whether the war was worth fighting existed, few denied that Iraqi control over Kuwait was contrary to the interests of the United States or the international community more generally, and critics of the war argued primarily that its costs would likely be unacceptably high, something that the Bush administration, with better information, was confident would not be the case. While Bush's popularity did rise significantly as a consequence of Desert Storm, there is no evidence that he chose to respond to Iraq because he believed that it would improve his hold on power, nor has there been any evidence of substantial misrepresentation of the situation in the Middle East or the likely outcome of a war.

Summary

The Persian Gulf War thus follows the classic pattern of a conflict driven by private information and divergent expectations. Saddam Hussein underestimated the both the resolve and the ability of the international community, and in particular the United States, to respond effectively to the Iraqi invasion. Conciliatory diplomacy prior to the invasion and an apparent pattern of casualty aversion gave Saddam the impression that the United States would not be willing to use force to reverse the conquest of Kuwait, and even if they were he doubted that the Saudis would permit Western forces to operate from their country. While some of these expectations were disabused over the course of the crisis, until the ground war Saddam remained convinced that the Iraqi army would impose high enough casualties on coalition forces to force a political retreat. It thus took a war to convince Saddam that his expectations were incorrect. Once the war began, however, the errors in his expectations rapidly became apparent, at which point he revised both his expectations and his demands downward, facilitating a quick political settlement.

THE ANGLO-IRANIAN WAR

More than a century before the Persian Gulf War, soldiers from the West found themselves in the area of the Persian Gulf fighting to preserve the autonomy of a small but important region. As with the Iraqi invasion of Kuwait, the crisis created by the Iranian (Persian) takeover of the Afghan city of Herat culminated in a short war that restored the initial status quo. Given the minimal consequences of the war—the deaths of a few thousand soldiers, but no political change—the war has left little imprint on the historical record.[36] While information on this conflict is limited, a clear picture nonetheless emerges of a war driven by overoptimism, in which the information gained through fighting facilitated a quick negotiated settlement, even in the face of great obstacles to negotiation.[37]

An Overview of a Little-Known Clash

The central stake in the war was the city of Herat, which, given the 1824 division of Afghanistan by three rival princes into the principalities of Herat, Kabul, and Kandahar, had been effectively independent for several decades. This situation presented an opportunity for Iran to reestablish control over Herat, which had belonged to Persia under the previous Safavid dynasty, and two separate shahs attempted to do so in 1838 and again in 1852. These incursions, however, inserted the country more centrally into the Great Game, the Anglo-Russian rivalry in Central Asia.[38] In that contest, the British attempted to ensure that local leaders in Afghanistan, through which an invasion of India would have to come, remained friendly; they were in particular opposed to an increase in Iranian control because of a belief that the Russians, who posed a significant threat to Iran, exerted excessive influence in Tehran. Thus, in both 1838 and 1852 the British responded to Iranian moves with a vigorous diplomatic effort that ultimately convinced the shah to withdraw his forces and agree to treaties repudiating any claim of sovereignty over Herat.

The issue returned to the fore in 1856, however, when an internal Afghan dispute led the prince of Herat to request Iranian aid. With this request as a pretext, the Iranians then moved in and, despite facing significant local opposition, established control over the city in October 1856. With negotiations between British and Iranian representatives in Constantinople dragging on, the British government in London authorized the governor-general in Bombay to declare war, which he duly did on November 1.

Subsequent military developments can be summarized quickly. Given the disasters that had accompanied previous incursions into Afghanistan—including an extended expedition from which only one British soldier escaped death or capture—the British were reluctant to intervene directly in Afghanistan and instead opted to land troops along the Persian Gulf coast. After doing so in December 1856, they seized the important trading center of Bushehr and then moved inland, where they defeated a significant Iranian force outside the town of Khushab in the largest battle of the war. Declining to pursue the defeated army inland, the British then moved up the Euphrates River, taking the city of Mohammerah (now Khorramshahr). On April 4, a few days after the fall of Mohammerah, plans for a significant incursion into the Iranian interior were interrupted by the arrival of news of a peace agreement signed in Paris a month earlier.[39]

Thus, less than four months passed from the British landing to the news of the peace agreement (and less than three from the start of fighting to the political settlement); while the British had had the best of it, at the point that the war ended, little fighting had occurred.[40] Nothing that the British had done directly compelled the Iranian army to withdraw from Herat. Why then did the shah agree to give up something that, as the repeated attempts over several decades to subvert Herat's independence clearly demonstrated, was quite important to him? The evidence in this case clearly indicates that the shah and his advisers underestimated British resolve—there existed a number of plausible reasons for them to do so—and thus adopted a policy that ran a risk of war with Britain were the British actually resolved, as they turned out to be. Consistent with the predictions of the informational mechanism, settlement came quickly, with the war ending basically as soon as the two sides could meet to negotiate and convey the peace agreement to the armies.

Private Information and Divergent Expectations

In contrast to the situation in many wars, the Iranians were not particularly optimistic about their military prospects in the event of a direct clash with the British.[41] They had reason, however, to doubt that such a clash would actually occur. While the British had established a clear interest in the fate of Herat (and the region more generally) in the past, several developments had provided reason to doubt the continued strength of that commitment.

The most important development was the annexation during the 1840s of the Sikh regions and the Punjab, which together provided a more natural and easily defensible northwest border for India.[42] Given this development, a reasonable observer might wonder whether the British would still fight for Herat, as the loss of Afghanistan to Russian control (no certain prospect) was now less threatening. Indeed, at the outset of a renewed Iranian attempt to annex Herat in 1851, Prime Minister Palmerston wrote to the ambassador in Tehran to say,

> You will, therefore, still endeavour to dissuade the Persian Government from advancing on Herat, and you may truly say that such a move would not be viewed with indifference by HM's Govt.; but you will be careful not to make any specific threat which HM's Govt. might not be disposed afterwards to carry into execution.[43]

While this communication obviously was not conveyed to the Iranians, it was a reasonable response to a changed strategic situation, which the Iranians undoubtedly had noticed. This particular incident ended with the shah pulling out of Herat, which the Iranians had successfully taken, in response to a forceful British remonstration. However, the British never actually mobilized troops, meaning that they did not actually demonstrate any willingness to spend blood and treasure to keep Herat independent. The shah might therefore still think that they would acquiesce in a seizure, especially if it was timed to coincide with British troubles elsewhere.

At the same time that the new frontier decreased Britain's incentive to fight for Herat, events in the disastrous First Afghan War—in which a British attempt to install a more amenable potentate in Kabul miscarried badly, with a significant British force utterly destroyed during an attempted retreat back to India—provided reason to expect the British to be reluctant to intervene directly in Afghan affairs.[44] Indeed, British

[172]

officials clearly were influenced by this precedent, with the governor-general in India (Lord Canning) sending a message back to London during the Herat crisis specifically to advise against any strategy that involved incursions into Afghanistan.[45] The British certainly still had tools at their disposal, as the invasion of the Gulf coast demonstrated, but those tools provided a less direct way of securing the independence of Herat.

Moreover, if the more secure border and the difficulty of intervening directly in Afghanistan provided a reason to think that when distracted the British might let the Iranians get away with seizing Herat, then the Crimean War appeared to provide the necessary distraction. While the British and their allies ultimately won the war, they did so only at high cost and in a less than completely decisive manner. Indeed, by some accounts far more was known in Asia about the Russian victory over the Turks at Kars than was known of the Russian defeat at Sevastopol; the Iranians apparently started their expedition with the belief that the British would have no spare troops to use against them.[46] Under these circumstances, Iranian decision makers could reasonably believe that the British might decide against diverting strength from a major conflict to prevent the seizure of a remote city whose strategic value had declined significantly. The end of the war with Russia removed this obstacle to British action, but unfortunately by the time that the Iranians realized their error and sought to correct it, events had proceeded too far, and they were unable to carry the negotiations through before the British, thinking that the Iranians would not back down without a fight, had committed to war.[47]

Thus, in October 1857, after the capture of Herat, the shah's primary adviser assured him that Britain would never resort to "coercive measures" and "would not move a soldier or a ship in connection with Herat."[48] It is thus unsurprising that the shah "appeared incredulous" when told that the British might occupy Bushehr, and that both he and his primary adviser expressed surprise on receiving the British declaration of war. The arrival of war, however, and especially the demonstration—through the occupation of Bushehr and the battle at Kushab—that the British were sufficiently resolved to fight forced a major revision in Iranian expectations. Initial clashes demonstrated that Iran lacked the manpower, money, or morale necessary to repel Britain; of the defeat at Kushab Amanat observes that "nothing could have more forcefully enhanced the image of British invincibility in the eyes of the demoralized Persian government." The revision in expectations is perhaps clearest in the shah's note to his representative in Istanbul: "Of course! Of course! Swiftly settle the matter in whatever manner you deem advisable. Do not allow it to come to severe hostility."

[173]

From this point, all that stood in the way of peace was the time that it took for communication to occur. Indeed, the governor-general in India, whom London had empowered to act, declared war ignorant of negotiations in Constantinople in which differences between the parties had narrowed.[49] After a brief delay when the news of the declaration arrived, negotiations moved to Paris, where an agreement was signed on March 4. The final agreement was quite close to the prewar British position, most obviously in Iran's agreement to withdraw from Herat, although they abandoned some tangential demands.[50] The extent of updating is clear, however, in the cheerfulness with which the shah greeted the news of the final terms, which although forcing him to back down on the central issues in the war at least permitted him to retain his throne.[51] The Iranians duly ratified the treaty and withdrew their army from Herat in short order.

To summarize, the Iranians saw a favorable opportunity to try to accomplish a fait accompli because they believed that when pushed the British might not prove willing to intervene. In this case, British resolve was private information, and because it was private (and not readily conveyed in a credible manner) it was possible, and indeed reasonable, for the Iranians to underestimate it. By contrast, there was relatively little divergence in expectations about the likely victor once war began. Thus, on the central issue leading to war, updating of beliefs occurred quite quickly, and the Iranians rapidly scaled back their ambitions, withdrawing from Herat (despite the fact that they had already captured the city and that the British lacked the means to force them out directly) based on the expectation that further fighting would only increase costs without improving the odds of political gains. These results thus are quite consistent with theoretical expectations: private information in this case could be effectively revealed only once the British demonstrated their willingness to fight, at which point the Iranians revised their expectations, lowered their political demands, and got out of the war, all in a relatively short period of time.

The Limits of Alternate Mechanisms

Neither of the other two primary mechanisms provides as convincing of an explanation for this case. A commitment problem explanation runs into the fundamental problem that no source that I have found records comments from anyone involved to the effect that war now would prevent a worse conflict at a later date. The British did wish to maintain the Afghan buffer against the possibility of Russian encroachment, but no one argued that the loss of Herat would put the Russians in a position to invade India or to demand concessions with respect to the subcontinent.

[174]

A more convincing interpretation of British policy is instead that they were interested in maintaining a clear reputation for protecting strategic interests in Central Asia—a view, given the centrality of private information to reputation, that is more consistent with an informational interpretation. Moreover, if one did believe that the war was motivated by a commitment problem, it would be hard to understand the course of events. Once the war began, the British did not directly relieve Herat, and the damage that they did to the Persians along the Gulf coast was hardly crippling. In other words, nothing in the military events on the battlefield would have resolved a significant commitment problem had it been contributing to the war.

At the same time, the case does demonstrate that mere suspicion of an opponent's intentions or reliability is hardly grounds for war without end. The Iranians had repeatedly violated agreements to respect the independence of Herat (most notably in 1838 and 1852) when they believed that they might be able to get away with doing so. Indeed, contemporary British sources consistently depict the Iranians as duplicitous and untrustworthy.[52] In response to this duplicity, however, the British did not launch a crusade to remake the Iranian polity; rather they acknowledged that they would need to make clear their willingness to intervene to the Iranians to deter further interventions. From this perspective, it is clear that simple mistrust associated with international anarchy is not sufficient to produce a major war.

Principal-agent dynamics likewise provide an unconvincing explanation for the war, although developments within the conflict provide some support for ancillary principal-agent hypotheses. In both Britain—at this point a partial democracy—and completely autocratic Iran, leaders had little to gain personally from fighting. In Iran, the Qajar Dynasty had developed an institutionally autocratic political system that gained legitimacy and internal cohesion from a system of effective negotiation and consensus.[53] Thus, despite (or because of) the existence of significant external threats, the shah did not have to worry greatly about his hold on power, at least so long as foreign powers like Britain did not attempt to unseat him. In Britain, Prime Minister Palmerston had gained broad esteem through his handling of the initially badly mismanaged Crimean War, and thus stood to gain little from further foreign adventures.[54]

That said, there is some indication that in fighting the war Palmerston—who could be described as an early advocate of liberal internationalism—was pursuing a policy that a substantial fraction of the British public did not endorse. News of the war reportedly was greeted in England "with a mixture of disgust and derision, and with anti-war demonstrations in Bradford and Newcastle"; the *Morning Star* referred to the war as a "Don Quixote campaign."[55] Moreover, the government was

[175]

also under fire for its conduct in the Second Opium War (aka, the Arrow War), which began concomitantly. Given this situation, Palmerston used, and most likely abused, some of the prerogatives of the prime minister. Parliament went out of session in July 1856, while the dispute in Persia was only just getting started, and only returned in February, by which time the war was well underway and there was little the opposition could do. Indeed, the government adopted the time-honored strategy of arguing that debate on an ongoing war would be contrary of the interests of the nation, arguing that it would be pointless to hold an open debate on the war without the relevant papers being available, but then refusing to release those papers because of the ongoing negotiations with Iran.[56] As a result, the opposition was limited in its criticisms to what could be gleaned from letters that soldiers sent home, which arrived long after the events they described had occurred. Nonetheless, the significant criticism that the government received was an indication of the limits to Palmerston's ability to engage in international adventures, especially when one considers that his government collapsed roughly contemporaneously over its handling of the Arrow War with China.[57]

Summary

As with the Persian Gulf War, the course of the Anglo-Iranian War is best explained by the informational mechanism. Although it was clear that the British would win in a military conflict, the Iranian government had reason to think that they might not be sufficiently resolved to fight one, given the stronger defense of India provided by the acquisition of the Sikh regions and the Punjab, British disinclination to involve itself directly in Afghanistan after the First Afghan War, and the apparent distraction provided by the Crimean War. In practice, however, this calculation proved to be incorrect: the Crimean distraction was gone by the time that the crisis over Herat had reached its peak, while the British found a military strategy that did not force them to fight in Afghanistan. Every indication is that the Iranians were prepared to back down as soon as it became apparent that the British were willing to fight—had faster communication between the adversaries been possible, it is unlikely that the war would have lasted more than a month, had it occurred at all. By contrast, the commitment problem mechanism appears to have been completely absent, while the principal-agent mechanism played at best a subsidiary role.

The Persian Gulf and Anglo-Iranian wars are not typical of the wars that scholars and policymakers tend to focus on, but they are typical of the wars that actually occur: both were short and not particularly deadly, at least by the standards of interstate wars, and both ended well before the loser's final military defeat, despite the potential for further escalation and a longer and deadlier war. In the Persian Gulf War, the coalition elected not to take advantage of initial victories to march on Baghdad; in the Anglo-Iranian War the politicians brought an end to the war just before the British force moved away from the coastline to launch a major incursion into the interior. In both cases, the best explanation for both the onset of the war and its short duration is provided by the informational mechanism. Leaders who initially miscalculated about their opponents' capabilities and resolve revised their expectations in response to the fighting and reduced their political demands accordingly, facilitating settlement. These findings thus accord with the prediction that informational wars, although prevalent, tend to be limited.

Equally important, the absence of significant commitment concerns in either case is consistent with the prediction that commitment problem wars will typically be more destructive. In neither case did leaders refer to concerns about shifting power in the way we observed in previous chapters, nor were there good reasons for them to worry about such concerns. This observation allays any potential concerns that commitment problem concerns are simply omnipresent in wars, in which case they would have provided a less convincing explanation for war destructiveness. Instead, it appears that these concerns are present only in the most severe of wars.

[7]

The Limits on Leaders

THE FALKLANDS WAR AND THE FRANCO-TURKISH WAR

Chapter 6 analyzed one kind of limited war, conflicts driven by the informational mechanism. This chapter examines a different kind, namely wars driven by principal-agent problems—in other words, misbehaving leaders—in domestic politics. In these wars, leaders adopt policies that are designed to serve their own interests rather than those of their constituents. I argue that these wars, like informational conflicts, are internally limited: the war will continue only so long as the leader is able to avoid censure, most likely by limiting the information available to opponents. As it becomes apparent that the leader is overstating the probability of victory and understating the costs of the war, opposition will grow and ultimately impinge on the leader's ability to continue the war. I further argued that in diversionary wars this process will typically occur quickly, because the very act of using the war to divert attention from other issues means that the leader will have difficulty lying effectively about how well it is going. By contrast, if the leader is simply pursuing a pet policy aim that the public does not share, she may be able to extend the war for a significant amount of time, but only if the costs remain relatively low. These wars thus may be long, but they will not be unusually deadly.

The case studies in this chapter are both ones in which domestic politics played a significant role in the onset, conduct, and termination of war. The better-known conflict is the 1982 Falklands War between Britain and Argentina, in which the Argentine military junta gambled that an invasion of the disputed Falkland/Malvinas Islands would improve their hold on power in the face of significant internal unrest. The lesser-known case is the 1919–21 Franco-Turkish War, in which the colonial faction in France hijacked foreign policy to pursue policy aims that did not serve broader French interests. The former case is a classic diversionary war, although a closer examination reveals the significance of other mechanisms. The latter fits the description of a "policy" war, in which downplaying coverage of events on the ground permits leaders to resist settlement so long as fighting is not particularly intense. In each case, the leaders ultimately found their freedom of action increasingly restricted by domestic opposition as the extent of their misrepresentations became clear; war ended in each case not because the leaders favored war termination but because they were left with no other choice.

THE FALKLANDS WAR

The Anglo-Argentine clash over the Falkland Islands in 1982 has been described as "the archetypal case of diversionary war."[1] According to this view, a military junta in Argentina that was steadily losing its hold on power distracted the public by launching an unexpected military adventure. This strategy succeeded in rallying the public behind the junta, at least until a British task force arrived and evicted the Argentines from their new conquests. Given this standard interpretation, this war provides a natural case for testing hypotheses about the principal-agent mechanism.

While it would be incorrect to ignore domestic politics, a closer look demonstrates that a purely domestic political explanation for the war would be woefully incomplete. The Argentines were willing to invade in large part because they underestimated British resolve and capability; thus, divergent expectations about the consequences of the use of force played a significant role in bringing about the conflict. Moreover, although temporarily successful, the diversionary strategy quickly backfired when the war went poorly for the Argentines. This case is thus consistent with the prediction that diversionary wars will typically be short.

History of the Conflict

The islands in question—known as the Falklands to the British and the Malvinas to the Argentines—lie several hundred miles off the coast of Argentina. Although of limited strategic significance—the famous Argentine novelist Jorge Luis Borges described the war as "a fight between two bald men over a comb"—the islands nonetheless have been the subject of disputes between several powers over the last few hundred years, including consistent Argentine claims over the entire period since the British established sole control in 1833.[2] Over time, the symbolic

[179]

significance of the islands increased, with the Argentines arguing that their transfer was necessary for the completion of decolonization, while the British argued that the principle of self-determination implied that the islanders—who wished to remain British—should be permitted to choose their own fate. British governments were to a significant degree entrapped, reluctant to anger the Argentines and entirely unwilling to pay the exorbitant costs that would be associated with funding an effective standing deterrent on the islands, but also unwilling to force the islanders, who had the support of a significant lobby in London, to acquiesce to a deal with which they were not comfortable.[3] They thus adopted a policy of equivocating on Argentine demands for a final resolution of the dispute, never directly rejecting Argentine claims but at the same time never allowing negotiations to progress toward a resolution. This strategy, however, grew more difficult over time, and when the Argentines began to push more aggressively for a final deal in the first few months of 1982 the British were forced to effectively reject Argentine demands.

An internal shuffle in the Argentine junta in December 1981 had produced a new, hard-line leadership that was predisposed to seek a final resolution to the Falklands issue by the end of 1982. The Argentines thus attempted to accelerate the negotiations while also planning for a possible invasion, scheduled if necessary to take place in the second half of the year.[4] In the event, however, the British refusal to make concessions on sovereignty combined with a sudden and unexpected crisis related to South Georgia—a polar island over which the two countries also contested sovereignty—led the Argentines to fear that the British were on the verge of reinforcing the islands, which if done would render an invasion impractical. As a result, the junta decided on March 26 to proceed with the military option. The invasion force departed on March 28; by the time that the British ascertained Argentine intentions on March 31 it was too late to do anything, and on April 2 the invasion succeeded with a minimum of fighting.

The conquest provoked strong reactions in both countries, with Argentines rallying to the junta while the British quickly assembled and dispatched a task force to retake the islands.[5] During the roughly three weeks that it took for the task force to reach the South Atlantic, the Argentines, working through a number of mediators, most prominently American secretary of state Alexander Haig, forwarded a variety of proposals for a negotiated settlement that would leave Argentina with final sovereignty over the islands while offering the British a variety of face-saving concessions. The British, however, resolutely demanded Argentine withdrawal as a precondition for negotiations, and thus no settlement was reached prior to the arrival of the task force.

[180]

Much of the war took place, unsurprisingly, at sea, where each side suffered losses, the biggest (and most controversial) of which was the sinking of the Argentine *General Belgrano*. After retaking South Georgia on April 25, the British landed troops at a secluded bay on East Falkland on May 21. These units then moved across the island, facing at times heavy resistance, to put themselves in a position by June 14 to compel the surrender of the forces in the island capital of Stanley. By this point, public opinion in Argentina had turned against the junta, which, after acknowledging its defeat, was rapidly turned out of power.

The Interaction of Domestic Politics and Divergent Expectations

How then can we explain the critical decisions in this case? Historical attention has focused primarily on the junta, as it both took the initial decision to invade the islands and failed to back down in the subsequent crisis period. That said, the British decision to retake the islands was not necessarily predictable prior to the event, and thus is also worth examining. The central questions for this case then are why, after decades of persistent negotiation, the Argentines took the gamble of invading the islands at this particular point, why the British decided to assume the risks and costs of a military response, why the two sides were unable to reach a negotiated settlement to the dispute prior to the British reconquest of the islands, and (an almost unrecognized question in the literature) why the war ended at that point.

Ultimately, both diversionary motives in Argentine domestic politics and divergent expectations about the likely consequence of a resort to arms are necessary to answer these questions. The domestic-political imperatives confronting the junta provided a reason for urgency, and thus ultimately prodded them to take risks that they would not have been willing to take had their hold on power been more secure. At the same time, the junta was willing to take these risks only because they underestimated how large they actually were, in line with the informational mechanism. Without overoptimism about the probability of a British military response and the likely outcome of a final military clash, the junta would not have been willing to invade in the first place.

Domestic Political Motivations

The military coup in 1976 that brought the junta to power was a response to widespread social unrest and poor economic performance.[6] It was thus welcomed by much of the country, especially among the more conservative parts of society that tended to share the political and social preferences of the generals. The new junta promised a neoliberal economic policy and a return to order, including the repression of leftists, that together would restore Argentina to wealth and stability. In practice, however, things did not work out as well as advertised. The repression deemed necessary to restore stability turned out to be remarkably widespread, with as many as thirty thousand Argentines "disappearing" in the dirty war in the first few years of military rule. Meanwhile, the economic reforms failed to generate the promised stability and growth, with continued high inflation coupled with increased unemployment and stagnant wages. Facing international criticism for human rights abuses and losing the support of parts of society that were frustrated by economic failures, the junta had to start to worry about fissures within the military that became more prominent over the course of 1981.[7]

As a result, pressure on the junta started to grow. In May 1981, General Videla, who had headed the military regime since the coup, completed his term in office; his successor in turn found himself unable to deal with the country's problems effectively and was ousted in December, precipitating a transition to hardliners under Leopoldo Galtieri. Starting with protests by the mothers of some of the "disappeared," Argentine civil society began to reassert itself, with leading political parties working together under the *Multipartidaria*, established in January 1982, to press the junta to restore civilian control of the government, while the trade unions began to demonstrate an increased willingness to protest over poor economic conditions.[8] In this context, the junta had to start to think more seriously about the possibility of transition. Indeed, several sources report that army General Galtieri aimed to use his time in office to put himself in a position to win an election after an eventual transition.[9]

Whether the goal was to stave off transition or to be in a position to fare well once transition arrived, the junta needed policy successes to enhance its legitimacy. Given the excesses of the dirty war and the economic difficulties, few obvious gains were available in domestic politics, and thus the natural solution was to look to the international realm. Here a contemporaneous territorial dispute with Chile over the Beagle Channel was headed for an unsatisfactory ending, with the junta anticipating that papal mediation would not yield a favorable conclusion. The Falklands dispute was little more promising, as the British—torn between the desire not to anger Argentina and a disinclination to force the Falkland islanders to join Argentina against their will—had effectively given the islanders a veto over any possible deal, but at the same time there were some indications that the British might not stand firm if pushed. Thus the new government decided to make acquiring sovereignty over the islands its primary goal for 1982, in the expectation that success in this venture would go a long way toward reestablishing the junta's reputation for effective leadership.[10]

[182]

In making this decision, the junta was not committing to a war with Britain, but it was demonstrating a willingness to consider an invasion if minimal demands were not met. On January 5, the junta formally ordered the military to develop plans for a bloodless invasion of the islands, while at the same time demanding that Britain agree to a new negotiating framework that would culminate in a transfer of sovereignty by the end of the year.[11] This diplomatic approach ran the risk of angering the British and of permanently alienating the islanders and hence hindering a deal, but the junta adjudged that risk to be worth taking. Indeed, when the initial talks did not yield the desired results, the junta broke one of the understood rules of past negotiations by publicizing the nature of the discussions, including the demand for sovereignty, thereby ensuring a diplomatic crisis (if still, from the British perspective, an eminently manageable one). At the same time, public comments hinted at the possible use of force. These diplomatic moves had the effect of heightening attention on the islands in Argentina, where newspaper editorials signaled to the public that it should be prepared for action of some sort on the issue.[12]

The risks that the junta was willing to run in the first few months of 1982 contrast sharply to the strategy that it had adopted in earlier negotiations, when it was willing to accept Britain's argument that quiet, low-pressure diplomacy that focused on minor issues while leaving more divisive issues like sovereignty for later had the best chance of eventually convincing the islanders to accept a transfer to Argentina.[13] In so doing, they recognized that they might alienate the islanders and thus make a transfer more difficult, but they gambled that if pushed the British would not be willing to suffer a diplomatic breakdown over a few thousand people whose continued support was a constant drain on state revenues. When that strategy failed, the decision to invade likewise was predicated on an acknowledged gamble that the British would be unwilling or unable to take back the islands. These gambles had been available earlier, but because of the risks associated with them successive Argentine governments had not seen them as worthwhile. Facing domestic political pressures, the junta thus was willing to take risks that previous governments had eschewed.

That said, there are limits to the principal-agent story. The most important one, discussed below, concerns the role of divergent expectations. It is also worth noting, however, that there are problems with the most extreme diversionary interpretation of the junta's behavior that emerged in the immediate aftermath of the war. Specifically, several commentators immediately after the war attributed the decision to invade to a desire to distract attention from large union protests over the state of the economy that began on March 30.[14] In reality, however, the order to invade was given on March 26, and the invasion force had left the harbor on March 28. Moreover, as was noted above, the invasion was the culmination of a broader strategy dating back several months, and thus could not have been tailored to immediate domestic concerns.[15] From this perspective, this case is consistent with the view that there are limits to degree of cynical diversion that elites can engage in.

Divergent Expectations and the Underestimation of Risk

More important, available evidence indicates that, even with its increased tolerance for risk arising out of a desire to enhance its domestic position, the junta was willing to authorize an invasion only because it believed that international opinion would be on Argentina's side, that the British would not fight back, and that any attempt against expectations to retake the islands would fail. These beliefs, although ultimately incorrect, were not without foundation, and it is thus worth investigating their origins and consequences in greater detail.[16]

Confidence that the international community would be amenable to the Argentine action arose from several sources. By framing the dispute in terms of decolonization, the Argentines had managed to get widespread support in the UN General Assembly, which had issued a number of favorable resolutions.[17] Meanwhile, the junta, which had impeccable anticommunist credentials, had established close relations with the Reagan administration in the United States, which led them to expect the United States to adopt a neutral position in the dispute.[18] While aware that the use of force might pose an issue, the Argentines hoped that a surprise invasion would allow them to take the islands bloodlessly; indeed, both the Falklands and South Georgia were occupied without British casualties, although the Argentine forces suffered some losses. In the event, however, the United States tilted ever more toward Britain over the course of the conflict, while the support that Argentina gained in the postcolonial world turned out to be of limited practical value.

The expectation that the British would not fight also had fairly solid foundations. Maintaining a true deterrent force on the islands, some eight thousand miles from Britain, would have been exorbitantly expensive and would have detracted from the far more important task of fulfilling Britain's responsibilities within NATO. As a result, the British had installed a platoon of marines—fewer than fifty men—backed up by the *Endurance*, an ice patrol ship of such limited capacity as to be "a military irrelevance."[19] This force was always intended as a trip wire (much like the NATO military presence in West Berlin), incapable of defending itself but generating a credible commitment to fight in the event of an invasion.[20] However, the British commitment to this force was never particularly strong, as evidenced by repeated debates over whether to keep paying for the *Endurance*, which culminated in the decision in June 1981 to withdraw the ship from service in 1982.[21] In this context, a bloodless Argentine takeover could easily be construed as providing the British with the opportunity to make a graceful exit from the scene.

Indeed, in previous negotiations, the British had encouraged the idea that they might want to make such an exit, telling their Argentine interlocutors that they would be perfectly happy to cede sovereignty if the islanders were willing to go along with the transfer.[22] The British raised no strategic or economic arguments for keeping the islands, and the Tory government had just two years previously been willing to negotiate about sovereignty behind the backs of the islanders, despite the risk that publicity would lead to attacks by the vocal and influential Falklands lobby.[23] The British refrained from making clear deterrent threats, both because they expected that invasion would only follow an extended coercive campaign and because they worried that any such threats might provoke the behavior that they were intended to prevent.[24] Moreover, several recent precedents gave the Argentines hope. The Thatcher government had been willing to negotiate away sovereignty over Rhodesia just the year before, while the Indian invasion of Goa in 1961 provided what the junta saw as a direct precedent: a military invasion of a colonial remnant that, while condemned in some quarters, ultimately was allowed to stand.[25]

Statements and actions at the time clearly demonstrated confidence that the British would not respond militarily. The invasion was carried out despite a significant lack of readiness for war, and the junta had no military plans for anything beyond the initial incursion.[26] The expectation was that the occupation forces would only be needed for a few months; indeed, the military began transferring soldiers back to the mainland almost immediately once conquest had been assured.[27] In an interview in the final days of the war, Galtieri observed that "though an English reaction was considered a possibility, we did not see it as a probability. Personally, I judged it scarcely possible and totally improbable."[28] Similarly, during the mediation of Alexander Haig, Foreign Minister Nicanor Costa Mendez commented that he was "truly surprised that the British will go to war for such a small problem as these few rocky islands."[29]

Moreover, the junta doubted Britain's ability to retake the islands even if an attempt were made. The Argentine military was geared toward the capabilities needed to control the South Atlantic, in contrast to British preparations for land war in Europe, and Britain would not be able to generate any sort of quantitative superiority in forces over the Argentines.[30] Initial clashes, in which the Argentines (erroneously) believed that they had imposed high costs on the British while thwarting an attempted landing, further encouraged the view that Britain could not retake the islands.[31] This view was not uniquely Argentine: given the tremendous distances involved and the lack of friendly bases from which to operate, many even among Britain's allies believed that the attempt was unlikely to succeed. Indeed, a number of scholars have argued subsequently that had the junta made more effective use of its available forces—instead of holding back its best units in interagency battles—it quite possibly could have won, and certainly would have imposed far greater losses on British forces.[32] Confidence in victory dovetailed with the view that the British would not fight, as even if the British valued the islands more than expected, it made no sense to attempt to retake them if that mission was doomed. Taken together, these were not unreasonable beliefs; the Soviets, for example, apparently also believed that Britain would not fight and could not retake the islands if it tried.[33] Of course, "so long as Argentina was unconvinced that Britain would actually fight or that, if it did, it would succeed there was no need to renounce its fundamental objective."[34]

The view from London, unsurprisingly, was different. Thatcher had little knowledge of military affairs, and thus relied on her advisers, such as Defense Minister John Nott, for guidance, although as a matter of principle she was inclined not to let the invasion stand. The military in turn quickly concluded that, although retaking the islands would not be easy—Admiral Fieldhouse at one point described it "the most difficult thing we have attempted since the Second World War"—it could be done at acceptable cost, at least so long as Britain moved quickly.[35] The navy's ability to dispatch a task force within a matter of days was a basis for confidence, with Nott literally overnight going from doubting that the islands could be retaken or that Britain could mount any sort of viable military response to believing "that a task force was a viable proposal and had a good chance of success."[36] Moreover, many in the Cabinet believed that a demonstration of British resolve in the form of the task force would convince the Argentines to back down, although this belief evaporated as successive attempts at mediation failed.[37]

Evaluating a Commitment Problem Story

Before moving on to discuss the way in which the conflict evolved over time, it will be useful briefly to evaluate the plausibility of a commitment problem interpretation of the war. Setting aside domestic political concerns, the junta had no reason to believe that time was not on its side. Militarily, the situation in the South Atlantic was likely to grow only more favorable to the Argentines over time: the British Ministry of

Defense had been trying for years to retire the *Endurance*, the only armed vessel in the region; the government also wished to withdraw the survey mission that constituted the only human presence on South Georgia. While both decisions were ultimately withdrawn prior to the invasion, it was clear that the solutions adopted were at best temporary fixes. Meanwhile, the British had decided not to withdraw the marines who were scheduled to be relieved in March, instead reinforcing them with their intended replacements.[38] In this context, the long-term trend seemed to be toward the invasion becoming even easier over time.

That said, as with the Persian Gulf War, there is some evidence that local concerns about shifting power may have influenced Argentine policy on the margins. The decision to invade if the British refused to negotiate seriously was made several months before the invasion, but at that point the junta did not intend to launch an invasion until later in 1982. The response in the British Parliament to the unexpected and unintended South Georgia crisis, most notably in its debate on March 23, raised fears that the British would reinforce the islands in the near future, thereby rendering it impossible to carry out the junta's strategy of a bloodless takeover.[39] The junta thus decided on March 26 to launch the invasion as soon as possible. This decision was militarily significant, in that it forced the junta to act with incompletely developed military plans and, more importantly, provided the British with a window to respond prior to the arrival of the stormy Southern Hemisphere winter. Shifting power concerns thus accelerated the war. By the time these concerns unfolded, however, the junta was already committed to invading, meaning that these concerns do not explain the decision to invade or, equally important, the manner in which the invasion was carried out. It is thus unsurprising that they did not play a significant role in the adversaries' subsequent conduct.

Views as the Conflict Unfolded

Intensive diplomacy followed the Argentine invasion, highlighted by mediation efforts by Haig, Peruvian president Belaúnde, and UN secretary general Pérez de Cuéllar. The historical record contains a remarkable amount of detail about the course of these negotiations, but the relevant developments can be summed up relatively quickly.[40] The Argentines had always recognized that negotiations would necessarily follow the invasion, and they were prepared to make a variety of what they saw as face-saving concessions to Britain to assure their success. That said, they resolutely insisted that any settlement reliably guarantee that Argentina ultimately receive sovereignty over the islands, although the formal transfer might occur as the outcome of some sort of negotiated transition process. The British, by contrast, never wavered in rejecting any approach that would reward Argentine aggression with sovereignty. It was this fundamental difference over the future disposition of the islands that prevented a negotiated agreement in the period prior to the insertion of the landing force.

During this period, it became possible for each side to at least in part revise its expectations about the consequences of the use of military force. Most obviously, the sending of the task force indicated that Britain was more willing to fight than the Argentines had believed, and the Argentines accordingly concluded that a war was indeed possible, although they continued to believe that there was a chance that Britain could be bought off at an acceptable price prior to actual clashes. Moreover, the response of other international actors provided an indication of how they likely would behave in the event of further escalation; in particular, the Argentines concluded that the Americans were providing significant assistance to Britain, indeed beyond what was actually given.[41] Overall, then, the junta grew increasingly desperate to avoid full-scale war as the crisis unfolded, although at the same time the military was confident that Britain would be unable to take the islands back.[42]

By this point, however, the junta had no real freedom to maneuver. The initial invasion had proven remarkably popular, with protests against the government transforming overnight into celebrations. Although quick to acknowledge that they would need to make some sort of concessions as a salve to British honor, the junta left no doubt that Argentina would retain sovereignty over the islands. Indeed, the escalating economic and diplomatic costs of the crisis meant that the junta was entirely dependent on success in the dispute for legitimacy, with the implication that rejecting concessions and continuing to fight was a "desperate gamble" that nonetheless was the only option that "held out any prospect of success."[43] Thus even relatively unambiguous evidence that Argentina would not manage to hold the islands did not bring about substantial concessions. Once the landing force was established, the best available option, if an increasingly unlikely one, was to stabilize the situation enough to ensure "a tolerable negotiating position."[44] Even at this point, however, the junta would not concede on the sovereignty issue, although they were less diplomatically exposed given that the British refused to contemplate any concessions once they were on the verge of retaking the islands by force.

After the local surrender on June 14, the Argentines lacked the capacity to insert a new invasion force on the islands. That said, an end to the war was not foreordained. In an interview just days before the surrender, Galtieri referenced the British evacuation from Dunkirk in 1940, which as he noted did not constitute a final defeat, and promised that not even the loss of the islands "would be the end of this conflict and our defeat."[45] The British had long worried that the Argentines might isolate the islands in a costly and painful diplomatic war of attrition; there is no reason why they could not have pursued a similar strategy of isolation and embargo after their initial gambit failed. Indeed, the British ambassador in the United States was reporting significant concern in Washington that retaking the islands would not necessarily bring about Argentine capitulation, and that the Argentines could carry out attacks indefinitely from the air, making the long-term cost of maintaining the islands unacceptable.[46] Had they done so, the British would have faced the uncomfortable prospect of an extended and costly campaign in the South Atlantic over a resource of tangential significance, with no real options for escalation that would not have brought about widespread international condemnation.[47]

Indeed, there is every indication that the junta contemplated exactly such a strategy. For several days the Argentines refused to recognize a British cease-fire declaration, while Galtieri called the population to the Plaza de Mayo in Buenos Aires to build support for a continued struggle of unspecified form. In practice, however, the continuation of the war was no longer an option. The disappearance of dissent had, in one scholar's words, "rested on only one point of support: the conviction that the war could be won," and with the surrender that support vanished.[48] The attempt to rally the public was met by chants of "the boys were killed; the chiefs sold them out," and the junta had to resort to violent repression of its own rally to restore order.[49] The junta had also lost legitimacy within its core constituency of the armed forces, where interservice recriminations quickly replaced the fight against the British.[50] With the loss of its ability to continue to prosecute the war, the junta had to concede defeat, accepting an end to the war and permitting a transfer within the military government to officers who had been uninvolved in the war.[51] Having invested their entire legitimacy in the acquisition of the islands, Galtieri and the junta had no incentive to end the war even after the garrison's surrender, but the public (and members of the military not in the junta), presented with incontrovertible evidence of incompetence and misrepresentation, refused to endorse such a move. These final events—little noticed in the diversionary literature—highlight the importance of constraints on the leadership's ability to continue to prosecute an ongoing war.

Summary

Overall, then, the decision to resort to force over the Falklands was motivated by both domestic politics and divergent expectations. The junta took a substantial risk in ordering the invasion, permanently alienating the islanders and ultimately provoking Britain into a military response for which the Argentines were unprepared. This willingness to take risks stemmed from an unambiguous domestic political incentive to pursue a policy success that would offset economic weakness, domestic repression, and diplomatic defeat in a separate high-profile territorial dispute with Chile. Yet the junta was willing to take this risk only because it underestimated its size. International opinion was far more hostile to the invasion than expected, with the United States providing Britain with extensive and unanticipated assistance. More important, the British turned out to be both more resolved and more capable than expected, willing to pay the costs associated with sending a task force to the South Atlantic to retake the islands in the face of resistance. Having successfully diverted opinion, the junta was unable to back down once it became clear that the British were willing to fight, and as a result no negotiated solution to the conflict was possible until after the Argentines had been forcibly evicted from the islands. That said, evidence of failure rapidly turned Argentine opinion against the junta, which found itself with no option but to end the war and surrender power.

This case thus highlights the limits of diversionary war, both as a strategy for embattled leaders and as an explanation for unusually extended conflicts.[52] The junta was able to use the Falklands issue for diversion, if only for a short period, only because it had recourse to a long-standing dispute on which most Argentines believed that their country had been eminently reasonable and had received nothing to show for their cooperation, and because their belief that the invasion was likely to succeed was quite plausible. Turning to the question of war duration, the failure of the junta's gamble rapidly became apparent, and as it did so its freedom to maneuver likewise vanished. The consequence was a war, but a war that never would have lasted more than a few battles, and that ended as soon as the Argentine public came to realize the deficiencies of their leaders' strategy. Indeed, it appears that the junta was prepared to continue the war after the fall of the islands, but by this point they were unable to do so. This is exactly the process anticipated in hypothesis 4a, which predicts that diversionary wars will typically be short.

THE FRANCO-TURKISH WAR

The 1919–21 Franco-Turkish War provides an example of what I refer to as a policy war. In this case, a French political faction that favored colonial expansion, salivating at the opportunities created by the partition of the Ottoman Empire, dragged the country into a war that the general public and the central government neither wanted nor knew much about. This interpretation helps to answer a number of questions about an otherwise puzzling conflict: why did France spend two years fighting a war against a country that it generally viewed favorably (and that had every incentive to find outside allies, given the range of foes stacked up against it), when the territory at stake was of little value and the political costs of the war were quite high? Not only can the principal-agent mechanism explain the decision to adopt a policy that did not serve broader French interests, it can also explain the decision of the colonial faction to stifle discussion of the war in France, which ultimately resulted in an extended, but not particularly costly, war.

History of the War

While the end of World War I on the Western Front was fairly orderly, postwar politics on the Eastern Front were extremely chaotic, with a host of new states jockeying for position in the power vacuum created by the collapse of the German, Austro-Hungarian, Russian, and Ottoman empires. In the years after the war, Eastern Europe and the lands of the former Ottoman Empire experienced conflict from the Baltic to the Holy Land. Even given the tendency to overlook this complex of wars, however, the Franco-Turkish War constitutes a particular lacuna in existing scholarship, overshadowed by the French actions in Syria and especially by the Greco-Turkish War.[53] That said, the Franco-Turkish War was an essential part of the broader Turkish independence struggle, while for the French it constituted an embarrassing defeat whose costs exacerbated the country's economic difficulties and contributed to the financial crises of the mid-1920s.

The war concerned possession of Cilicia (roughly corresponding to modern-day Çukurova), a region in southern Anatolia that lies to the north of Syria.[54] Inter-Allied negotiation during World War I about the postwar disposition of the Ottoman Empire resulted in an agreement to allow France to claim mandates over Syria (including Lebanon) and Cilicia, while the British acquired Mesopotamia, the Italians acquired a region in southern Anatolia west of Cilicia, the Greeks took Smyrna (modern Izmir) in western Anatolia, and the Armenians were allotted an independent national state in the East. The basic nature of the secret Sykes-Picot Agreement, which in May 1916 formalized the British and French claims on Ottoman Territory, can be seen in figure 7.1 below: the blue zone of direct French control included coastal Syria and much of southeastern Anatolia, although in practice the British began impinging on the French zones almost from the outset (for example, quickly gaining unilateral control over Palestine) and the French never established an effective presence in the interior regions.[55]

[191]

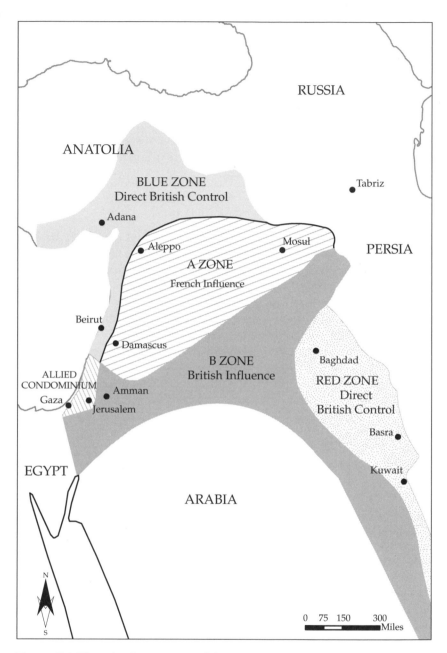

FIGURE 7.1 Plans for the partition of the Ottoman Empire, including the zones claimed by France

French policy throughout this period involved repeated retreats to second-best options. What the French really wanted in Anatolia was a return to their pre–World War status quo, in which they had exercised a preeminent political and economic influence in the Ottoman Empire.[56] In this goal, the French were undercut by their allies, however, first with the March 1915 Russian demand for Constantinople, to which the British assented, and later with expanding British territorial interests in the Middle East.[57] The French thus retreated to trying to secure at least a generally accepted claim to some portion of Ottoman territory, centering on Syria, yet even the British concessions in the Sykes-Picot Agreement were undermined as the British, who had far more soldiers in the Middle East, gradually encroached on French claims while encouraging undesired Greek activism in western Anatolia.[58]

Meanwhile, the political situation in Turkey was changing quickly, as the supine Ottoman Porte—the traditional government of the Ottoman Empire—lost legitimacy to a nationalist resistance headed by Mustafa Kemal (later venerated as the Atatürk, or father of Turks) and located in the Anatolian interior.[59] As the extent of Allied demands on Turkey became clear, the Porte's strategy of relying on British magnanimity to retain power became increasingly unpopular, while increased organization and early military successes encouraged the view that the Turks could successfully oppose their enemies.

The emergence of Kemal's force coincided with the initial guerrilla resistance against the French, who had assumed control over Cilicia in November 1919 and almost immediately encountered low-level but persistent military opposition.[60] A successful uprising at Maraş in early 1920 spurred further opposition throughout the region, forcing the French to consolidate their overstretched forces in a few strategic locales. Unable to defeat the resistance, the French tried to reach a negotiated agreement with Kemal, but their refusal to contemplate withdrawal from Cilicia and their support for the Porte ultimately prevented any durable agreement.[61] In the face of continued guerrilla resistance, the French were able to achieve some local military victories, but never came close to pacifying the contested region. Ultimately, with the impossibility of overcoming the resistance (at least at acceptable cost) clear, and with growing threats to the more highly valued possessions in Syria, the French negotiated a humiliating withdrawal in the Ankara Agreement of October 1921, whose terms were later carried over into the 1923 Treaty of Lausanne.

The war thus ended with the French withdrawing from Cilicia in humiliation, having abandoned almost all their claims to the region. Their troubles in Cilicia were an embarrassment internationally and had undermined efforts to achieve political goals elsewhere.[62] Their withdrawal was also a disaster for the native Christian population, largely Armenians, who, having already suffered mass killings during World War I, generally fled rather than await retribution for their involvement in French occupation.[63] For the Turks, the victory was a significant step on the way to establishing a truly independent Turkey, free of the constraints that had hindered the Ottoman Empire, although full achievement of the Turkish national state came only with victory over the Greeks and the Treaty of Lausanne.

Domestic Politics

Ultimately, a convincing explanation for the Franco-Turkish War must examine the role of the French colonial party in pursuing goals that, for all their talk of French prestige, were not in the interest of the country as a whole.[64] French colonialists had long been concerned with expanding the French Empire, despite the absence of clear national interests beyond simple prestige that would be served by doing so. In particular, oft-cited economic benefits had little basis in reality, and business leaders were correspondingly uninterested in acquiring new colonies. Moreover, colonial aspirations served as a distraction from France's central security concern, which was Germany. In the Cilician case, the aspirations of the colonial party created conflict with Britain at precisely the time when the French needed a close relationship with the British for protection against Germany. The colonialists recognized that conflict over Cilicia would be unpopular and thus took steps to misrepresent their actions, the ease of their task, and the level of resistance that they faced. This strategy generated systematic misperceptions in Paris, which in turn ironically tied the hands of the colonial faction when it finally decided that the time had come to abandon Cilicia. It was these actions that ultimately produced both an unnecessary (and unnecessarily extended) war and a substantial humiliation for France.

The French Colonial Party and the Interests of France

The French colonial party was relatively small in numbers but exercised a significant influence over French foreign policy in the decades prior to World War I. Adherents to the party were spread across some fifty overlapping societies, with a total membership that certainly numbered under ten thousand and most likely numbered under five thousand.[65] Starting around 1890, however, they took advantage of their strength among civil servants and the chronic instability of French governments under the Third Republic to exert a consistent influence on French foreign policy in favor of colonial expansion.[66] Their influence further increased during World War I, when the urgency of the war on the Western Front distracted their opponents from more distant issues and made "the abdication by the cabinet of its responsibility for colonial war aims even more complete than it might otherwise have been."[67] Thus, for example, François Georges-Picot, who represented the French in the negotiations with the British that led to the Sykes-Picot Agreement, drafted his own instructions for the negotiations, which highly optimistically called for France to receive the entire Mediterranean coast from Egypt to Cilicia, extending inward to Mosul and the Tigris; Prime Minister Briand accepted them without amendment.[68]

In pursuing the acquisition of colonies, the colonial party was consistently hampered by the difficulty it had identifying and articulating positive benefits that would ensue from an expansion of the empire. The British had already secured the most valuable colonies, most notably India, while the zones that the French acquired were frequently of no real economic value. Indeed, for many of the French colonialists, value for French finance and industry was hardly relevant: the motivation was French prestige.[69] Under these circumstances, it is unsurprising that French business figures were frequently less excited about colonial acquisitions than members of the colonial party wished for them to be. Thus, for example, during World War I colonialists in government formed a commission with the sole intent of recommending colonial expansion and called in leading business figures in the expectation that they would provide support for such an enterprise. They were thus sorely disappointed when business leaders expressed almost no interest in overseas acquisitions and instead called for increased tariff protection.[70] It is thus hard to argue that France's colonial acquisitions served a useful economic purpose.[71]

More important, colonial aspirations if anything detracted from French security. Unlike Britain, which benefited from the defense provided by the English Channel, the French were a continental power that had to worry about threats from both Germany and Italy. The Germans in particular were a serious concern: they had imposed a humiliating defeat on France in the Franco-Prussian War and came close to doing so again in 1914 in a war that the French won narrowly and only with the help of a number of significant allies. At a general level, then, pursuing imperial dreams in the Levant was a distraction from the far more important task of finding a way to keep Germany down and to begin the process of domestic recovery from the war.[72] Moreover, the French were keenly aware that security against Germany in the long run would require assistance from allies. Communism in Russia, isolationism in the United States, and general weakness among the remaining Western powers left Britain the most attractive option in this regard and thus provided good reason to avoid antagonizing the British.[73] For the colonial party, however, the defeat of Germany left Britain as the only other significant imperial power and thus as the natural opponent, even while the two sides were formally allied during World War I.[74] The inevitable consequence was to risk alienating the very country on which the French relied for security over overseas interests of minimal strategic significance. Indeed, colonial disagreements contributed notably to the deterioration in Franco-British relations in the years after the war.[75]

Given the costs of acquiring and administering new colonies, the absence of clear economic benefits, and the geopolitical dangers of estrangement from Britain, it is unsurprising that the colonial faction lacked popular support. Indeed, the colonialists were quite aware of this unpopularity, as when a leading colonialist noted that "French public opinion will for a long time preserve its desire to repress all efforts at colonial competition. . . . This attitude is unfortunate but incontestable."[76]

Colonial Policy in Cilicia: Hiding the War

Given the unpopularity of their preferred policy, the colonial party responded by pursuing their desired ends while going to great lengths to avoid attracting the attention of Paris. They thus maintained expansive territorial demands in Cilicia while simultaneously doing everything they could not to call on resources (particularly soldiers) from metropolitan France. Instead, they responded to the insufficiency of French forces in the Levant by drawing on Muslim colonial troops (despite fears that they might prove unreliable in fight against Muslim Turks) and, more problematically, by co-opting local Armenians to serve as legionnaires, despite their new soldiers' manifest deficiencies and proclivity for attacking Turkish civilians.[77] The upshot of these strategies was to leave French forces perpetually undermanned and staffed with unreliable troops, whose repeated attacks on the local population ultimately helped to ensure the failure of French efforts to control and pacify the region.

The colonialists further protected themselves from constraints imposed by Paris by using their levers of control to ensure that as little information as possible concerning Cilicia, and especially concerning the difficulties that French forces encountered there, reached the eyes and ears of parliamentarians and the public in Paris. The colonialists had a significant informational advantage to begin with—the French public tended to confuse Cilicia with Silesia (which lies in present-day Poland), for example—and thus needed only to prevent unambiguously discrediting information from appearing while muddying the waters in public discussions.[78] To the end of the war, colonialists in the Quai d'Orsay—the French foreign ministry—imposed a general news blackout that, despite early Parliamentary complaints that the government was hiding the truth from the people, prevented the opposition from mounting effective criticism.[79] The effectiveness of the news blackout—the socialist press ultimately was limited to reprinting reports from British and German newspapers outside the control of the colonialists—meant that critics had to resort to arguments about the illegitimacy of colonialism, to which the colonialists could respond with equally principled arguments about protecting the local Christian population.[80] The evidence for more powerful critiques about the failure of pacification attempts and the likelihood of backlash against the local Christians, which would more clearly have undercut the colonialists' case, was simply unavailable.

Public Opposition, Aborted Withdrawal, and Painful Defeat

At the same time that hiding the war allowed the colonial party to pursue goals that most Frenchmen would not have supported, it also did much to ensure their ultimate failure. The reluctance to ask for resources from the metropole meant that General Gouraud, the French commander in Syria and Cilicia, would be perpetually short on men and unable to pacify the region directly; reliance on the Armenians, while addressing the manpower shortage, ultimately worsened the problem by alienating the local population. Meanwhile, even those limited funds that government was providing for action in the Levant came under threat as parliamentarians became increasingly aware that the government was hiding something.[81] By March 1921, Prime Minister Briand had to deal with a parliament that was "slashing his military budget by hundreds of millions of francs every day."[82] These developments are illustrative of the constraints that domestic opposition can place on policymakers intent on adopting policies that do not serve the national interest. In practice, these domestic constraints, combined with complications in Cilicia and Syria discussed below, tied the hands of the colonial party sufficiently to convince them the attempt to control Cilicia was now a lost cause.

As a result, General Gouraud and the colonialists prepared a change of strategy. The colonialists had always valued Syria more highly than Cilicia, and the emergence of a significant threat from Arab nationalists under Emir Faisal committed to incorporating Syria into a broader Arab state posed a serious threat that ultimately induced the French to forcibly seize Damascus in July 1920.[83] Even after the victory over Faisal, who decamped to become king of Iraq, the French faced continuing unrest in Syria, which culminated in a significant revolt between 1925 and 1927.[84] Given this situation, the French thus had a strong incentive to reach an understanding with the Kemalists that would permit the redeployment of troops from Cilicia to Syria. Kemal was aware of French

[197]

difficulties, however, and consistently rejected proposed deals that would formally recognize Turkish sovereignty over Cilicia while leaving the French with effective control. By September 1920, therefore, the colonialists reluctantly concluded that Cilicia was not worth the effort required to hold it.[85]

At this point, however, the colonialists found themselves entrapped by their own chicanery. Precisely because of their efforts to suppress negative information about events in the Levant, policymakers in Paris were uninformed about the extent of difficulties there. As a result, Prime Minister Georges Leygues, who admitted to an Armenian diplomat that he knew almost nothing about events on the ground in Cilicia, reversed the withdrawal policies and ordered a significant escalation of the war.[86] This move torpedoed any chance at a settlement in the immediate future but did nothing to improve the French position in Cilicia, as the troops devoted to the escalation in Cilicia were needed back in Syria in short order to restore a deteriorating situation there. It was thus only when Gouraud, who obviously was fully aware of the problems in the region, in consultations in Paris demanded full withdrawal from Cilicia that the French returned to seriously (and increasingly importunately) seeking a negotiated settlement.[87] This development, while not anticipated in the theoretical discussion of domestic politics and war, nonetheless is consistent with the observation that to be effective in diverting policy, agents must keep their actions relatively quiet.

To summarize, a small but well-placed faction in French politics was able to divert policy to pursue its preferred aim of colonial expansion, despite the high costs and limited benefits of such a policy for France more generally. By limiting discussion of the war, the colonialists ensured that they could continue to stake a claim to Cilicia even in the face of sustained guerrilla resistance. The strategies that they used to maintain that policy ultimately undercut their ability to win the war, however, and the combination of increasing checks domestically and growing threats in the Levant ultimately forced them to back down. That said, a reasonable case can be made that this strategy would have worked better at a different time. The First World War meant that the French public was unusually attuned to foreign policy and, given war weariness, more reluctant to pay significant costs for minor policy aims.[88] In other settings, the colonialists might have been able to entrap Paris by first staking a claim and then convincing the government to provide the resources to protect a claim that it would not originally have advanced. In such a situation, the war might have been substantially longer than it actually was.[89]

The Limits of Alternate Mechanisms

In contrast to the clear domestic political story, neither commitment problems nor overoptimism seem to have played a significant role in this war. While the reestablishment of control over Anatolia certainly strengthened Turkey relative to its neighbors (including the French in Syria), the French were not worried about this development, largely because they expected to be able to restore their historically good relations with the Turks once the war was over. As for optimism, while overoptimistic prognostications were present, they are better understood as part of a duplicitous campaign to justify the French claim to Cilicia than as genuine beliefs about the likely outcome of a war.

Consider first potential concerns related to the restoration of Turkish power. For several reasons, the French did not feel threatened by this accretion in capabilities. Given continued disputes with the Greeks, British, and Armenians, in the immediate term Kemal and the Turks had every reason not to intervene in Syria, the one place where they could potentially have threatened French interests. Indeed, during the war Kemal had generally been cool to suggestions of military cooperation from Arabs in Syria, given the danger that such cooperation would complicate attempts to end the war in Cilicia and hence delay the restoration of the generally friendly relations that Turkey had had with France in the past.[90] Thus the French appear not to have worried that surrendering Cilicia would simply have allowed the Turks to bring the war to Syria.

In the longer term, of course, a Turkey that successfully fought off its opponents (still not guaranteed at the time of the Ankara Agreement) might have been able to threaten France's hold over Syria. That said, there were several reasons why the French were not worried about such gains.[91] The French had had close relations with the Ottoman Empire prior to World War I and, as was noted above, would have preferred a situation in which they retained a significant degree of influence over a unified Turkey to its complete partition. A not insignificant benefit was that the Turks might be induced to repay the Ottoman Empire's debts.[92] Moreover, given the historical rivalry between the Ottoman Empire and Russia, a united Turkey was a logical partner in the attempt to prevent the spread of communism, a particular concern of French governments during this period.[93] Most important, the significant animosity felt by Arabs in the Middle East toward the Turks—a consequence of hundreds of years of repressive rule—meant that the obstacles to reestablishing the Ottoman Empire were in practice insurmountable, a point that Kemal appears to have been fully aware of. It is thus unsurprising that debates about withdrawal from Cilicia focused on the fate of the Christian population in Anatolia and on possible reputational costs in France's Muslim colonies; indeed, French parliamentarians concluded that that an agreement with Turkey would provide better security for Syria than occupation of Cilicia possibly could.[94]

An argument grounded in overoptimism about the ease of pacifying Cilicia is similarly unconvincing. This is not to say that overoptimistic prognostications were not present: colonialists predicted both that France's civilizing mission and experience with prior Muslim colonies would lead the population to accept them and also (contradictorily) that the local population in Cilicia consisted primarily of Christian Armenians who would constitute logical allies.[95] On closer analysis, however, these predictions appear to have been less than entirely sincere. Thus, for example, the colonial party heavily lobbied local Muslims and Christian Arabs to submit petitions for a French protectorate that they then used as evidence of spontaneous local support; similarly, the overestimation of the size of the Armenian population relative to the Turks was accomplished in part through the deliberate deception of omitting from the count of Turks anyone who was not ethnically Turkish, who did not adhere to Sunni Islam, or who spoke Arabic, despite the fact that the vast majority of those so omitted could be counted on to side with the Turks in the event of a conflict.[96] To the extent that the French government in Paris was overoptimistic, it was as a consequence of these sorts of misrepresentations.

Moreover, in many respects the French were unusually accurate in their assessments. They were the first of the World War I allies to recognize the importance of Kemal's movement, negotiating with him even before fighting broke out.[97] As a result, the obsequiousness of the Porte provided little basis for comfort. Likewise, it was almost immediately apparent that the Armenians—who first opposed the French presence altogether out of a desire to construct a Greater Armenia and then repeatedly attacked Turkish civilians in retribution for past attacks or, later, to undermine negotiations with Kemal—would constitute difficult and counterproductive allies.[98] Given the quick difficulties that the French encountered and their unwillingness to contemplate the sort of escalation that would have been necessary to address these problems, it is hard to argue that the war lasted for as long as it did because of sustained French overoptimism.

Summary

In summary, to understand what happened in the Franco-Turkish War, we need to examine developments within French domestic politics. In the face of a well-organized colonial movement, the French government lost control over imperial policymaking, not only overseas but in Paris itself.[99] As the primary group that cared about imperial policy, the colonial faction in France secured the relevant policymaking posts and then proceeded to pursue policies that few in France would have chosen. The expansion into the Levant, including Cilicia—despite the almost complete absence of interest among the business and finance figures who were supposed to benefit, the strong public demand for demobilization, and the likelihood that this policy would damage relations with key allies and thus undermine French security in Europe—thus involved the colonialists making political demands on Turkey that the French public was not willing to back up. Under these circumstances, when fighting ensued, the French forces in the Levant were incapable of asserting control over the situation. The colonialists nonetheless remained committed to trying to salvage some sort of position in Cilicia, with the result that their begrudging concessions were never enough to match what the Kemalists knew they could get by just extending the war until discontent in Paris forced a unilateral withdrawal. The threat to Syria, combined with the constraints increasingly imposed from Paris, prompted Gouraud to begin a unilateral withdrawal from Cilicia, but here the colonialists were undermined by their own strategy, as a new prime minister, unaware of the extent of the difficulties in Cilicia precisely because of the strategy adopted by the colonialists, overruled the decision to withdraw, thus substantially extending the war before a final French withdrawal in 1921.

Overall, then, this war almost certainly would not have happened had the colonial party been kept on a tighter leash, and it certainly would have ended sooner. These findings are consistent with the argument that in "policy" wars—nondiversionary principal-agent conflicts—leaders will attempt to limit the availability of information, and that to the extent that they are successful in doing so the war may be relatively long. Indeed, had the colonialists not chosen a particularly inauspicious time for their overseas adventure, it is likely that the center would not have imposed constraints as quickly and hence that the war might have dragged on substantially longer.

Neither the Falklands War nor the Franco-Turkish War can be explained without recourse to domestic politics. In the former case, a government with a shaky hold on power saw a military clash as an opportunity to restore its position, and thus took a gamble that it otherwise would not have undertaken. In the latter, a small set of policymakers with interests at odds with those of the broader public whom they theoretically represented fought a war to advance claims that their constituents would not have backed. In each case, these decisions brought about the deaths of soldiers in the pursuit of objectives that can hardly be seen as advancing their interests and that in any event were not achieved.

The only silver lining to this story is that in each case fewer soldiers died than the responsible policymakers might have been willing to sacrifice. In the Falklands War, an initially successful gamble backfired as the British proved both more willing and better able to respond militarily than the Argentine junta predicted; by the end, reaction on the streets and within the armed forces meant that any attempt to extend the war beyond the fall of Stanley—a military possibility, if not a political one—was doomed to failure. In the Franco-Turkish War, the knowledge that they would not have the backing of Paris forced the colonial party into a strategy that limited the direct costs of the war and militated against substantial escalation, albeit at the cost of lowering the probability of victory. Both wars were thus internally limited: the Falklands conflict ended before the Argentine leadership would have chosen to end it, while the war in Cilicia was less deadly than it likely would have been had the colonial party had free rein to spend France's resources as it wished. At the same time, commitment problems played no role in either of these wars, consistent with the argument that they tend to produce wars more destructive than either of these.

Conclusion

Most interstate wars are limited, in either duration or intensity. In a small number, however, intense fighting continues for years without the two sides resolving their differences. Indeed, in rare cases one side in a war categorically refuses to negotiate with its opponent, despite the extraordinary costs of a war to the death. The long, intense conflicts, which I have referred to as "unlimited wars," may be rare, but they are responsible for most of the suffering caused by war in the past two centuries. Existing scholarly explanations for these wars have not been entirely satisfying, however. Political scientists and historians have generally been more interested in why wars start than why they end and frequently focus exclusively on the most destructive and politically consequential conflicts. These studies have contributed many valuable insights, but they have also left an important gap—the field simply did not have a convincing answer to the question of what separates the many more limited conflicts from the few that are particularly destructive. Why, in short, do the two sides in a war sometimes come to a quick resolution of their dispute, while in others they refuse to resolve their political disagreements even in the face of extended and painful fighting? This book, I hope, helps to fill this gap in our knowledge.

In this conclusion, I first recapitulate the arguments and findings in this book. The next section highlights their implications for two related topics—civil war and strategies of conflict management—that are not discussed elsewhere in the book. The final section discusses the implications of my findings for questions about the future of war and of international politics more generally.

[203]

RECAPITULATIONS: LITTLE WARS, LONG WARS, LARGE WARS, AND WARS TO THE DEATH

In this book, I argue that the size of wars is a function of the causal mechanisms that drive them. When countries fight because their leaders disagree about what is likely to happen once the shooting begins, events in the war cause expectations to converge—more quickly when fighting is more intense—until a mutually agreeable settlement appears. When leaders pursue private goals at the expense of the national interest (the principal-agent mechanism), they face domestic constraints that force them either to settle quickly or to ensure that fighting is not too intense, so that they can distract public attention from the war. Only commitment problem wars are thus not logically limited. In situational commitment problems, a declining power begins a preventive war to prevent the anticipated consequences of decline; doing so, however, entails high war aims that the declining power will be reluctant to relinquish even in the face of battlefield difficulties. Moreover, in some cases, the target of a preventive attack concludes that its adversary is dispositionally aggressive and undeterred by the costs of fighting, and hence that peace can only be guaranteed by fundamental change in the adversary's political system. Given these beliefs, negotiation is futile; the only acceptable form of war termination is the opponent's unconditional surrender. The remainder of this section recapitulates the logic of these claims and the evidence adduced to support them.

Informational Wars

Under the informational mechanism, war participants fight because they disagree about what is likely to happen should they resort to war. Indeed, as Blainey noted, they generally disagree quite substantially, expecting not only to win but to do so quickly and at low cost. Once fighting begins, optimistic expectations will be challenged by events on the battlefield. If both sides expect to win the opening battle, at least one will be surprised when it is fought. That surprise in turn provides reason to revisit one's expectations, allowing for an increased probability that the war will end badly. Once this possibility is acknowledged, political settlements that previously seemed unattractive will become more palatable. As the participants lower their demands, a settlement that both prefer to continued fighting will eventually appear; at this point, the participants should identify the settlement and end the war. Indeed, because both sides expect to win quickly, the amount of surprise in the war will be large, with the result that updating of beliefs (and hence settlement) will occur in a matter of months rather than years, especially when fighting is relatively intense.

Both quantitative and qualitative evidence provide support for these claims. In statistical analysis, high war intensity, which should proxy for the speed at which information is revealed, is associated with shorter wars and with quicker settlement, although demonstrating this effect required separating out the commitment problem wars, which can be both long and intense. Two case studies provide further support. In the nineteenth-century Anglo-Iranian War, an initial disagreement about British resolve was cleared up by the onset of fighting, allowing for a quick political settlement. More recently, the Persian Gulf War of 1991 occurred because Saddam Hussein underestimated both American willingness to fight for Kuwait and the opposing coalition's ability to evict Iraqi forces from occupied Kuwait at acceptable cost; once the course of fighting demonstrated the coalition's resolve and capability, Saddam backtracked dramatically in his political demands, thus permitting a settlement.

By contrast, arguments based on divergent expectations provide an unconvincing explanation for larger wars like World War II. Hitler and his generals recognized that Germany's initial expansion was quite risky and had the potential to plunge the country into an unaffordable long war, but they undertook it anyway; even more glaringly, they refused to contemplate negotiation with the Soviet Union even as military defeat and occupation became ever more glaringly inevitable. Similarly, despite the undeniable military disaster of the collapse of France and the expectation of an imminent German invasion of the British Isles, British leaders refused altogether to negotiate with Germany, even when German demands proved to be relatively moderate. While the belief that the German economy was vulnerable provided some (poorly grounded) basis for confidence, the informational mechanism simply cannot explain why a country that had seen its strategy for victory completely undone and that faced the imminent threat of total defeat would have been unwilling to even consider negotiation.

Taken together, these findings indicate strongly that wars driven primarily by the informational mechanism tend to be limited. As most wars are limited, this finding is quite consistent with the view that differing expectations about how the war will go account for most uses of force. That said, this mechanism cannot account for the most destructive wars.

Principal-Agent Wars

Domestic-political explanations for war are unified under what I refer to as the principal-agent mechanism. Under this mechanism, leaders use war to pursue goals, be they a stronger hold on power or policy goals in foreign or domestic politics, that the public would not support, at least at the costs that the war entails. They must worry, however, about the constraints—whether they be institutional checks on policy, the army's refusal to fight, or the possibility of removal from power—that society can place on a ruler who is seen as following undesirable policies. As a result, highly visible diversionary strategies, as with the Falklands War, will be successful at best for a short period of time before the public catches on that the leaders are running unnecessary risks or refusing reasonable settlement offers and withdraws support. If the leader is able to keep the war less visible, however, most obviously by ensuring that the fighting is not particularly intense, then she will have greater freedom of action and likely will be able to extend a war for longer if necessary. In either case, however, the constraints placed on leaders mean that these wars will not be both long and intense.

Again, both qualitative and quantitative work provides evidence in support of these propositions. As with the informational mechanism, the path to peace depends on the revelation of information through war, in this case to the public. The relationship between war intensity and the speed of settlement in the noncommitment problem wars thus is also consistent with theoretical expectations for the principal-agent mechanism. Moreover, consistent with the argument that constraints on leaders limit their ability to impose policy unilaterally, I find that leaders of democratic states—who face greater institutional constraints on their actions—tend to settle more quickly, although the relevant relationships are less statistically robust than some project. Contrary to some prior work, I also found no evidence to suggest that partially democratic war losers present particular obstacles to settlement.

Case studies of the Falklands War and the Franco-Turkish War are similarly consistent with these arguments. In the Falklands War, a military junta that faced threats to its hold on power was willing to run risks that it would not otherwise has countenanced, if only because it underestimated the size of those risks. By invading the Falklands, the junta brought about a war with Britain in which it could not make concessions on the central issue—final sovereignty over the islands—without losing power, and thus it was willing to continue the war even after the Argentine garrison was forced to capitulate. Military failures, however, turned the public against the leadership and left it unable to continue the war or even to maintain its hold on power. The post–World War I Franco-Turkish War by contrast was undertaken not to divert the public from domestic troubles but because foreign policy had come under the control of a faction that wished to expand France's colonial empire, even in the face of general opposition to such a policy and at the cost of a significant deterioration in France's security in Europe. In line with theoretical expectations, the colonial faction stifled coverage of the war in Paris while taking steps to limit demands on the central government and thereby increase their freedom of action. This approach allowed for an extended war, although in the end funding cuts imposed by a hostile legislature combined with threats to their position in Syria to force the colonialists to abandon their claim to Cilicia, the territory in dispute with Turkey. Once again, therefore, domestic constraints on undesired leadership behavior brought about war termination, although because the colonial faction was able to keep its actions quiet this war lasted substantially longer. Overall, then, both theory and evidence indicate that principal-agent problems in domestic politics can account both for relatively short wars and for longer conflicts, but that they cannot account for wars that are both long and intense.

Commitment Problems: Preventive Wars and Unconditional Surrender

Only commitment problems, in their two different guises, remain to account for the most destructive unlimited wars. When faced with a situational commitment problem, declining powers fight preventive wars because they fear that if current trends continue they will be forced into unpalatable concessions or a war on worse terms. War can only address this problem if it forestalls the decline, however, and typically the victory necessary to prevent the decline from occurring will be quite large. As a result, in wars driven by this mechanism, the initiator typically has unusually large war aims and is unwilling to settle on intermediate terms, a combination that frequently leads to unusually risky military and diplomatic strategies. If the initial attack miscarries, which, given the risky strategies often chosen, it frequently will, the result can be a quite extended military conflict, in which even significant suffering may not be sufficient to convince the declining power to settle. I argue that these wars will be rare, precisely because these wars tend to be risky and costly, but that when they happen they are likely to be unusually destructive.

This argument has a number of testable implications for the conduct and termination of war. Statistical tests reveal that larger shifts in relative prewar capabilities, which proxy for anticipated future shifts, are associated with more destructive wars, an in particular with increased difficulty reaching a political settlement. By contrast, when leaders have reason to fear decline, they are more likely to resort to relatively risky strategies, with the result that these wars are unusually likely to end through the conquest of one side by the other, sometimes quite quickly.

[207]

Qualitative evidence also provides significant support. The nineteenth-century Paraguayan War, which almost totally destroyed Paraguay while imposing incredible costs on its opponents as well, started because Paraguayan president Francisco Solano López feared that a rapprochement between Argentina and Brazil and an effectively joint intervention in Uruguay would put his neighbors in a position to partition his country. Facing this increasing threat, he decided that an aggressive war to break up the incipient alliance and restore the government of Uruguay was preferable to the risks associated with accepting continued relative decline. He thus launched an aggressive and risky war that, although understandable given the difficult situation he faced, ultimately miscarried badly. Adolf Hitler in Germany similarly believed his nation to be headed for a serious decline that, if not addressed, might well lead to the extinction of the German race, although in this case his beliefs were grounded more in his ideology than in the structural situation of Germany. Moreover, by the late 1930s, he had concluded that the combination of German rearmament and Stalin's purge of the military had opened up a narrow window of opportunity for Germany to make the substantial territorial gains necessary to address the broader decline identified by his ideology. He thus launched an incredibly aggressive and risky war with the ultimate goal of annexing much of European Russia; as long as the war continued he resisted any form of settlement with the Soviet Union. Shorter case studies of the Crimean War, the Pacific War in World War II, and the Iran-Iraq War provide further support. In the Crimean War, the British believed the Russians to be on the verge of acquiring control over Constantinople and the Black Sea Straits and responded by fighting for war aims designed to drive the Russians back to a point at which such gains were impossible. The Japanese believed that only significant expansion could maintain their status as a great power; the urgency with which they sought that expansion increased dramatically as they perceived a window of opportunity created by World War II in Europe and as an American oil embargo threatened to throttle their military forces. Finally, Saddam Hussein saw the disruption created by the Iranian Revolution as a golden opportunity to revise the results of past disagreements in Iraq's favor and potentially eliminate the threat that an ideologically hostile regime posed to his hold on power.

Wars to the Death

Even in preventive wars, the declining power remains at least theoretically open to negotiation, if not on terms that the rising power would be at all likely to accept. Thus, once his attack had clearly miscarried, Paraguay's López demonstrated at least a potential willingness to talk, while Hitler was willing to negotiate with Britain and seems to have expected that some form of rump Russia would survive beyond the Urals. In both these wars, however, as well as in the Pacific component of World War II and the Iran-Iraq War, some participants categorically refused to negotiate with the existing regime on the opposing side, instead preferring to fight a war to the death. This behavior, which I refer to as a sincere demand for unconditional surrender, guarantees a particularly long and bloody war; the preference for such a war over any possible settlement is puzzling from a theoretical perspective. Existing work, however, has not provided a good explanation for it, nor can any of the three mechanisms discussed here provide a fully convincing account.

I argue that sincere demands for unconditional surrender arise out of a two-part process. First, a declining power launches an aggressive preventive war designed to forestall that decline; then the target of the preventive war attributes the attack not to the fear of decline but to the innately aggressive disposition of the opposing ruler or of the opposing society more generally. This inference, however, implies the existence of what could be called a dispositional commitment problem: an innately aggressive opponent will view any peace deal as simply providing an opportunity to choose the optimal time and place for the next attack. Having attributed the war to the enemy's aggressive character, the target of the war thus logically concludes that a viable peace will require the fundamental remaking of the opposing political system, something that can only be achieved following a total victory. The result is a war to the death. I further argue that this process is more likely when the rising power that is targeted in the attack lacks the intentions attributed to it by the power that fears decline. This discussion identifies a range of testable hypotheses that can be tested in case study analysis.

The case studies examined here correspond closely to this argument. Thus, in the Paraguayan War, López feared that Brazil intended to work with Argentina to partition Paraguay. This fear was reasonable, but Brazilian actions after the war, when they had every opportunity to extinguish Paraguayan independence, demonstrate that they were misplaced; the subsequent Brazilian refusal to negotiate—justified on the basis of López's iniquity—thus conforms to the argument that innocent targets will be more likely to demand unconditional surrender. Moreover, the Argentines, who gave every indication of wishing to annex Paraguay, did not join in Brazil's refusal to negotiate, consistent with the argument that targets who are in fact hostile will understand the motivation behind the attack and thus be open to talks. Similarly, in World War II, none of the Allies shared Hitler's theory of world politics, and thus they did not intend to do what he feared, nor did they understand the motivations behind his actions. As a result, they attributed his actions to an innate desire for war and concluded that no settlement with Hitler or, for most Allied leaders, with any potential German representative would bring a sustained peace. Moreover, my argument is able to account for more microlevel features of this conflict—such as the willingness to relax the terms of unconditional surrender for Germany's allies, the way in which Allied leaders' views of Hitler changed over time, and the particular emphasis that American leaders placed on unconditional surrender—that are also consistent with my argument. Similarly, the Japanese attack in World War II was grounded in fears that the United States intended to render Japan a third-rate power, quite possibly through war in the next few years, that were inconsistent with American intentions. The Iranian refusal to negotiate in the Iran-Iraq War also can be read in this light, although definitive evidence in this case is unavailable. By contrast, the case study of the Crimean War, in which Britain's preventive actions did not induce Russia to refuse to negotiate, is also consistent with the argument about innocence, as the Russians clearly wished to acquire Constantinople, the development that the British were fighting to prevent.

Taken together, then, this process provides the logic for unconditional surrender: an aggressive action, taken out of perceived necessity, is seen as evidence of an aggressive disposition that must be expunged. This, in other words, is how people end up in a war to the death.

Implications: Internal Conflict and External Intervention

The primary goal of this book has been to develop a better understanding of the determinants of particularly long and bloody wars. To render the analysis more tractable, I focused on interstate wars, setting civil wars aside. A reasonable question is to what extent the findings from this work can be translated to intrastate conflicts. Because civil wars differ structurally from interstate wars in certain significant respects, direct translation of results is not possible. Nonetheless, there are several relevant implications of this study for our understanding of civil wars; this section thus highlights those implications. Similarly, by focusing primarily on the decisions of the leaders of the warring parties, I set aside the question of how outsiders might facilitate or encourage war termination. That said, because effective strategies will directly address the underlying causes of war, a better understanding of why wars continue or end provides a useful basis for generating policy prescriptions. The second part of this section thus highlights several implications for policy that follow from the findings here.

[210]

What Lessons for Civil Wars?

This book focuses on interstate wars, which include most of the particularly deadly wars throughout history. From a theoretical perspective, interstate wars are easier to understand because participants are generally more easily separable—the participants typically are not competing over a population whose preferences lie somewhere between those of the principal actors—and the spoiler problem associated with splintering actors poses much less of an obstacle to settlement.[1] For both these reasons, interstate wars are closer to the assumptions of the standard bargaining model and thus provide a more attractive initial testing ground. That said, given the frequency of civil wars, both today and likely in the future, it is appropriate to assess the implications of the findings here for our understanding of internal conflict.

The central finding here, that preventive wars arising from shifting power make for the worst wars, dovetails with an emerging consensus in the civil wars literature.[2] Absent secession, a comparatively rare outcome, a prerequisite for a functioning state following civil wars is the demobilization and integration of the armed forces on each side. As Walter notes, however, to the extent that one side cheats in the demobilization phase or manages to secure control over the unified forces after demobilization, the other side will find itself in an extremely dangerous position and will certainly not be able to enforce the political agreement on which the two sides settled the war.[3] Given the prevalence of this problem in civil wars, it is thus unsurprising that they frequently prove far more intractable than interstate wars and thus that civil wars last substantially longer than interstate wars on average. That said, while these wars tend to be long, the limited capabilities of weak states and especially of many rebel groups mean that fighting is frequently less intense than in the interstate context.

Other differences between civil and interstate wars are starker. A now large literature highlights the role of natural resources in generating intractable civil conflicts.[4] In many ways, natural resource wars are comparable to principal-agent conflicts, in that the wealth that accrues from controlling an exploitable resource tends to accumulate in the hands of a small number of leaders. An important question in these conflicts is how these leaders are able to convince the people whom they claim to represent to follow them. To the extent that outsiders are able to convince rebel soldiers that the leaders for whom they are fighting, who often grow rich without allowing the benefits of wealth to trickle down beyond the highest echelons of their movement, are the only ones who would really suffer from the imposition of peace, these leaders likely will find it far more difficult to reject settlement on terms the government proposes.

[211]

Where civil wars seem to differ most clearly from interstate wars is in the difficulty of clearly identifying the relevant actors. As was noted above, spoiler processes can lead wars to continue even when political leaders of all relevant factions have concluded that a proposed settlement is preferable to continued fighting. Perhaps more problematically, in failed states, such as Somalia since 1991, it may be the case that it is impossible to identify an appropriate set of actors with whom one can conduct negotiations. The difficulty that the United States faced in identifying a strategy to end the Iraq insurgency arose in part because it faced opposition from a wide range of domestic factions, many as antagonistic toward each other as they are toward the Americans. This situation is not one that can be captured easily using any of the mechanisms analyzed here, meaning that it is not possible to derive clear lessons from this study for these types of cases.

Ending Wars Once They Have Started

Postmortems on violent conflict frequently bemoan the failure of the international community to act prior to the outbreak of violence, when it is presumed that the costs of intervention would have been low and the positive effects in securing peace would have been great.[5] In an ideal world, policymakers would be able to identify potential violent conflicts before they begin and intervene to prevent their occurrence. Yet achieving this goal will likely remain elusive: there are strong theoretical reasons why identifying future wars out of the larger set of potential conflicts is difficult, and there is no indication that the international community would be willing to commit sufficient resources to make proactive measures in all countries in which conflict might occur possible.[6] In this context, then, it makes sense to discuss strategies for ending wars after they have begun.

One of the central implications of the bargaining model of war, and one of the key findings of this study, is that wars are characterized by equifinality: there are multiple individually sufficient causes of conflict. If we grant the argument that the problems that led to war must be resolved before we can return to peace, then any discussion of strategies for ending wars must acknowledge that appropriate policies may vary based on the underlying causes. An intervention strategy designed to address preventive motivations for war will not be effective if the underlying problem is simply that both sides think that they are likely to win on the battlefield. We must also be cognizant of the ways in which different problems interact, as with the observation that in diversionary wars the informational mechanism is also frequently active.

[212]

Existing work on strategies for ending ongoing wars has only just begun this process, however. Most theoretical studies of mediation, for example, focus on only one mechanism, and few scholars explicitly address multiple mechanisms or mechanisms in combination. Kydd's work, which constitutes an important exception to this generalization, highlights the ways in which prescriptions may change as the underlying problem leading to conflict changes.[7] When the problem is private information, Kydd finds that only biased mediators can credibly convey information (as pessimistic statements from the perspective of the side that the mediator favors can convince that side to make concessions). In contrast, he finds that an unbiased mediator will generally be more effective when the participants wonder whether it is wise to trust the other side, an endemic concern in the context of preventive motivations for war. These observations thus demonstrate the importance of emphasizing the contingent nature of policy prescriptions and—a step that the field has shown little interest in undertaking—of developing tools to help policymakers identify the underlying causes of conflict so that they can better identify which strategies will likely prove more effective.

This book has been primarily concerned with evaluating the relative effectiveness of different mechanisms in explaining wars of varying destructiveness, and thus comments on policy prescriptions will necessarily be preliminary. That said, several significant implications emerge from the findings here. In general, ending informational wars requires convincing participants of their overoptimism and thus inducing them to update their beliefs. Doing so may be difficult, but especially better-informed mediators, such as representatives of the great powers, may be able to credibly convey information that will convince participants to modify their expectations and hence their demands. Ending wars in which principal-agent dynamics are at play will require alerting the citizens of their leader's shenanigans and providing other assistance so that they can impose greater constraints on that leader's ability to continue the war. Ending preventive wars will generally be most difficult, and will require more active intervention, for example through the formation of defensive alliances that will provide that declining power with some guarantee that it will not suffer excessively from its decline. The unfortunate implication here is that outsiders will probably have the greatest leverage over those conflicts that are most likely to end quickly without outside intervention, while they will have the least control over precisely those wars that they will most want to influence.

To turn to more specific observations, several nonobvious implications for ending wars can be derived from this framework. Thus, for example, highly public bargaining may be counterproductive when the primary concern is private information (as the two sides may have incentives to adopt extreme positions to signal resolve) but efficacious when leaders on one side are extending wars unnecessarily (as knowledge of the other side's bargaining stance may help the constituents of leaders who are pursuing a diversionary strategy to recognize this tactic and respond appropriately). Similarly, biased military intervention will frequently be extremely effective in bringing an end to informational conflicts (indeed, often the source of disagreement is precisely whether an outsider will intervene), but it may do little to bring about war termination when fighting is driven by other mechanisms. One general implication of this book is thus that work on strategies of conflict management will benefit from careful consideration of the underlying mechanism producing violence.

PROGNOSTICATIONS AND TERMINATIONS

With the end of the Cold War and the collapse of the Soviet Union, the apparent end of ideological conflict and a consequent convergence in worldwide preferences provided a basis for forecasts of the substantial reduction or even end of the violent conflict that has dogged humanity throughout its history: pessimists were fighting a rearguard action based on predictions that many saw as fanciful at the time and that have as yet failed to materialize.[8] Violent conflict in the Balkans and genocide in Rwanda provided initial correctives to these expectations; September 11 and the subsequent wars in Afghanistan and Iraq demonstrated that ideological conflict persists and that interstate wars are still possible. Even so, however, many people still believe that major wars, or interstate wars more generally, are rapidly becoming obsolete.[9] It is certainly true that the Western and Central Europeans, long the protagonists of many of the biggest interstate wars, have devised a political system in the European Union in which resort to violence seems unthinkable. Elsewhere in the world, however, violence remains more than just thinkable, while the likely retreat of US hegemony in the next few decades may make the world a more unpredictable place.

Indeed, the apparently inexorable rise of China has raised concerns about a possible war with the United States in the future. China's rise generates understandable concerns among Americans about the concessions that the United States will need to make in the future in light of Chinese power.[10] Indeed, concern about the implications of China's rise has led its neighbors to seek to ensure continued American involvement in Asia, while some in the United States have called for a more confrontational approach based on mistrust of long-run Chinese intentions.[11]

[214]

That said, there are several good reasons to doubt that China's rise will lead to great power war. Indeed, precisely because such a war would be enormous, the potential participants have a very good reason to avoid it; it is likely only to the extent that the declining power—in this case the United States—can identify a reasonable plan for preventing the decline from occurring. While the United States is and will for the foreseeable future remain the dominant military power in the world, there are significant limits on its ability to impose its will directly on China. China's economic growth, while driven by the coasts, is broad-based and is driven by factors—most notably an extremely large population—that could not be changed by anything short of a war to break up the state of China, an undertaking that no serious academic or policymaker has contemplated. To put it simply, it is entirely unclear how, having started such a war, the United States might plausibly expect to be able to end it: at a minimum, it would need to break off several of the large and economically prosperous coastal cities like Shanghai, but having done so it would have no plausible policies for what to do with them. In such a context, even someone who is genuinely concerned by the significant military buildup that China has undertaken will lack a convincing argument for why war would serve American interests.

The most likely source of conflict between China and the United States would be a dispute over Taiwan. Theory here indicates that the Chinese, as the rising power, should see little need to demand immediate concessions on this issue, as they can wait until later, when they can negotiate from a position of greater strength. Indeed, Chinese policy on Taiwan appears to be following this prediction quite closely: the military buildup around the Taiwan Straits appears to be intended primarily to deter the Taiwanese from making any overt move toward independence. Moreover, while Taiwan is symbolically and economically significant, its broader strategic significance is limited: in contrast to British fears about Russian control over Constantinople freeing the Russians to move into the Mediterranean, there is little more that China could do after acquiring Taiwan that it could not do beforehand. War could nonetheless still occur if, for example, the Chinese leadership underestimates American resolve, or possibly if threats to their rule induced the Chinese leadership to launch a diversionary invasion.[12] Given the size of the combatants, such a war would no doubt be destructive, but it would not be the unlimited conflict that has been the greatest concern of commentators. It is more likely that China and the United States will manage the transition without war, with the United States making some tacit concessions along the way.

While the implications of this study's findings for the US-China relationship thus are a source for optimism, some of the more general results provide reason for caution. Great power war in the nuclear era is admittedly unlikely. Indeed, given the reduction over time in the number of great powers, there simply are fewer dyads in which such a conflict could occur. That said, one point emphasized in this book is that one does not need great powers to have big wars. The Paraguayan War and the Iran-Iraq conflict provide evidence that medium powers are perfectly capable of inflicting horrendous costs on each other. Moreover, while these wars too are rare, there is no trend to indicate that they are decreasing substantially over time: after all, the Iran-Iraq War was hardly in the distant past.

And if nuclear-armed states are unlikely to fight each other, the process of acquiring nuclear weapons can be profoundly destabilizing. While nuclear threats are frequently incredible, given that any attack will provoke an equally destructive response, a nuclear-armed state nonetheless protects itself against bullying by existing nuclear powers and opens up a range of options for conventional mischief making.[13] Thus the shift in relative capabilities associated with the acquisition of nuclear weapons provides a potent rationale for preventive war, the more so because destroying a nuclear weapons program—which will be reliant on a limited number of nuclear plants and testing centers—may seem easier than preventing other sources of shifts in relative capabilities. From this perspective, the American attack on Iraq may be only the first of several such preventive wars. Moreover, as nuclear proliferators learn from past targeted attacks on reactors in Iraq (in 1981) and Syria (in 2007) to disperse and harden potential targets, opponents will discover that truly eliminating a nuclear program will require deep penetration by ground forces, as the international coalition did in Iraq in 2003. This motivation for preventive war is unlikely to disappear soon.

Separately, the dramatic changes in the international system with the end of the Cold War may have both good and bad implications for the nature of war in the future. On the one hand, the end to superpower competition eliminated the imminent threat that any conflict might escalate into a catastrophic great power war, and also decreased the likelihood that peripheral conflicts will serve as proxy wars in which external assistance sustained each side's military capacity and hence extended the potential amount of bloodletting. On the other hand, it is unclear that superpower competition necessarily made for worse wars. Precisely because of the possibility that a local war might escalate to include one or both of the superpowers, with the specter of nuclear use that such a possibility raised, leaders in both the United States and the Soviet Union had a strong incentive to keep third-world conflicts limited. Indeed, frequently the worst conflicts arise when no one is able or willing to address the fears of declining powers. In the Paraguayan War, outside powers simply did not care enough about events in the interior of the South American continent to intervene with any enthusiasm. In the Iran-Iraq War, an essential element of the conflict was the mistrust with which the Revolutionary regime in Tehran viewed the rest of the world, which complicated any outsiders' attempts to bring about peace. In World War II, the powers were completely sorted into the two camps by the end of 1941, leaving no outsiders who could intervene to allay the concerns of the participants. From this perspective, then, the reduced interest of the great powers in events in the developing world may in fact remove a check on big wars, possibly making such conflicts more likely.

To say that such wars are as likely to occur as they were in the past is not, of course, to say that they are imminent. The most deadly wars remain for very good reasons thankfully rare; it is thus entirely possible that we could see several decades pass without a particularly large interstate war. Because declining powers can, and frequently do, respond to their decline by simply accepting a loss of influence, predicting the occurrence of preventive wars is extremely difficult: we may be able to identify situations, as with the US-China relationship, in which relative decline makes such conflicts a possibility, but it is much harder to pick out from among the dyads in which conflict is likely those in which war will actually occur. When it does, however, as it almost certainly will, a better understanding of its dynamics may help policymakers to find the strategies that will help bring fighting to a close.

Notes

INTRODUCTION

1. Wawro (1996, 42, 282–283).

2. See Lieber (2007) for a contrary view of initial expectations that builds on recent historiography.

3. On the potential for escalation in the 1929 conflict, see Lee (1983, 99) and Wei (1956, ch. 5). For the Iranian decision to adopt maximalist war aims, see for example Bulloch and Morris (1989, 11–12, 103) and Seifzadeh (1997).

4. This figure relies on data introduced in chapter 2. A war is coded as high-intensity for this figure if battle deaths over time are found above the seventy-fifth percentile in the data either in absolute terms or when adjusting for the total population of war participants. While these thresholds are of course rather arbitrary, a similar picture would emerge from any reasonable division. Likewise, while data availability forces me to use only battle deaths here and elsewhere in this project, inclusion of civilian deaths or military deaths outside battle would not affect the point that a small number of wars are responsible for a disproportionate share of deaths.

5. See Stanley and Sawyer (2009) for a similar discussion of equifinality and war.

6. Fearon (1995).

7. Clausewitz ([1832] 1976, 75–76).

8. Realists have also disagreed about the underlying mechanism driving great power wars, with Waltz (1979, 170–172) attributing such wars primarily to miscalculation, while Copeland (2000) attributes them to fear of decline.

9. On the war between France and Austria, see King (1967, vol. 2) and Blumberg (1990). For the Soviet-Japanese clashes, see Coox (1977) and Coox (1985).

10. For a recent estimation of total Paraguayan deaths, see Whigham and Potthast (1999).

11. Studies that examine war duration statistically actually go back quite some time: see Weiss (1963), Horvath (1968), Morrison and Schmittlein (1980), and Vuchinich and Teachman (1993). These studies were limited by existing computing power in their ability to include covariates, however, meaning that most assessed only the appropriateness of different assumptions about the functional form of the relationship between elapsed time or total fatalities and the probability of war termination.

More recent studies have been able to include a more extensive list of more interesting covariates.

12. Bennett and Stam (1996).

13. Richardson (1960), Cederman (2003).

14. This review is necessarily cursory and omits a number of important studies on which I build in the next chapter. Even the most pertinent studies have tended not to examine the question of what separates the most destructive interstate wars from those that are either shorter or fought less intensely, however.

15. For particularly important exceptions, see Goemans (2000) and Reiter (2009).

16. Chapter 2 contains a more detailed discussion of the various measures of war destructiveness.

1. EXPLANATIONS FOR LIMITED AND UNLIMITED WARS

1. For the paradigmatic presentation of the bargaining model, see Fearon (1995). Powell (2002) and Reiter (2003) provide useful, if now slightly dated, reviews of the bargaining model literature. See also Jackson and Morelli (2009).

2. Clausewitz ([1832] 1976, 87).

3. Blainey (1988). For studies that build directly on this mechanism, see Smith (1998a), Gartzke (1999), Filson and Werner (2002), Reed (2003), Slantchev (2003b), Powell (1999, 2004), and Smith and Stam (2004).

4. Thucydides (1952). For additional work that points to a connection between shifting power and war, see for example Organski (1968), Levy (1987), Copeland (2000), and Powell (2006).

5. Kant ([1795] 1957). For recent examples, see Levy (1989), Snyder (1991), Goemans (2000), and Mansfield and Snyder (2005).

6. Fearon (1995, 382). More recently, Powell (2006) has noted that from a formal perspective issue indivisibilities are simply another example of the commitment problem mechanism that lies behind preventive wars driven by shifting power. Rather than fighting, each side would prefer an *ex ante* gamble (e.g., a weighted coin flip) in which their probability of getting the issue at stake was tied to their probability of victory in war; the problem is that neither side can credibly commit not to resort to war if it loses the coin flip. This logic gets to the commitment problem in a different way from the shifting power logic, however.

7. For arguments that indivisible issues are important, see for example Kirshner (2000) and Toft (2006). For an analogous argument about the limited importance of indivisible issues in war, including a number of additional examples of such limitations, see Reiter (2009, 47–50).

8. Hassner (2004), Goddard (2006).

9. For other problems with risk acceptance as a variable in international politics, see O'Neill (2001).

10. See Slantchev (2003a) for the logic of this mechanism.

11. Powell (1999) and Coe (2011) present the logic of this argument. This mechanism could potentially account for the war proneness of seventeenth-century Sweden, which was relatively poor and sparsely populated and hence could not afford to

support a large army on its territory for long. As a result, whenever the threat of war was high enough to force the Swedes to gather their army, they needed to invade one of their neighbors almost immediately to permit their army to requisition supplies on enemy territory. Given that it severely antagonized all of Sweden's neighbors, this strategy worked remarkably well until the disaster of Poltava and subsequent defeat in the Great Northern War cast Sweden from the ranks of the major powers. See Frost (2000, esp. 116–118, 200–208). Coe argues that this mechanism is also responsible for several more recent conflicts, most notably the 2003 invasion of Iraq, although his account depends heavily on both commitment problems (associated with the potential Iraqi acquisition of nuclear weapons) and overoptimism about the likely costs of war. I am unaware of cases in which this mechanism played a primary role in bringing about war.

12. Thucydides (1952, 49), Copeland (2000).

13. For power transition theory, see Organski (1968), Organski and Kugler (1980), and Kugler and Lemke (1996). See also Gilpin (1981) and Kennedy (1987). For additional quantitative tests of power transition hypotheses, see Houweling and Siccama (1988), Kim and Morrow (1992), de Soysa et al. (1997), and Lemke (2002). For a more general discussion of theories of general war, most building on the role of shifting power, see Levy (1985).

14. Fearon (1995), Powell (2006).

15. *New York Times*, "Bush Officials Say the Time Has Come for Action on Iraq," September 9, 2002, A1; Woodward (2004, 202).

16. Trachtenberg (1991), Trachtenberg (2005).

17. Fearon (1995).

18. See for example Gilpin (1980) and Kennedy (1987).

19. That said, Stalin did believe the Soviet Union to be on the rise, following Marxist-Leninist ideology about the inevitable replacement of capitalism with socialism. Theory, however, indicated to him that the capitalist states were dominated by those who would lose most heavily from this transition, and thus that reassurance was unlikely to work. He thus expected the Soviet Union to be attacked by the West. See for example Roberts (1991, 18). Nonetheless, he ordered consistent compliance with the terms of his pact with Germany even as it became more apparent that Germany was planning to attack; the Germans were fully aware that the Soviets were doing everything that they could to signal benign intentions. Gorodetsky (1999); Chief of Council for the Prosecution of Axis Criminality, *Nazi Conspiracy and Aggression*, vol. 6 (1946), 997–1000.

20. *New York Times*, "Iraqi Dictator Told of Fearing Iran More than He Did U.S.," July 3, 2009, A10.

21. Fearon (1995), Powell (1999, ch. 4).

22. Springhall (2001, esp. 210). See also Rock (1989, ch. 2) for a discussion of a similar British decision not to oppose the American rise in the second half of the nineteenth century.

23. On internal balancing, see Waltz (1979, 168). There is also the danger that such production may exacerbate the security dilemma and thus bring about competition that would not otherwise have occurred.

24. Several studies, starting with Fearon (1995), have discussed the reasons why this particular agreement is not generally viable.

25. More imaginative diplomatic options may have been available historically, as when Godfrey of Bouillon, the leader of one of the largest armies headed for the Holy Land on the First Crusade, left his brother Baldwin and Baldwin's family as hostages of the Hungarian king as a pledge of good behavior during the period Godfrey's army traversed Hungary and hence was in a position to wreak considerable disorder, as indeed a number of crusading armies did both previously and subsequently (Runciman 1951, esp. 148). In the modern era, the use of formal hostages has of course ceased to be a tool of diplomacy.

26. McCrum (1978).

27. Wagner (2000).

28. Reiter (2009, ch. 9) points however to additional commitment concerns that motivated the Germans to continue the war on the Western Front even after Brest-Litovsk, which explain the continuation of the war overall. Ultimately, defeat on the Western Front resulted in the abrogation of Brest-Litovsk, with the affected regions either gaining independence or reverting to the control of the Soviet Union.

29. Leventoglu and Slantchev (2007).

30. The Iraqi civil war that followed the invasion ultimately entailed many years of additional fighting for American soldiers, of course. This fighting was motivated by a fundamentally different set of concerns, primarily relating to sectarian struggles for control of the Iraqi government, from the concerns that motivated the conventional phase. It is entirely conceivable that with a better plan and a bigger occupation force the coalition might have avoided this violence; in any event, the preventive motivations for the initial invasion were addressed by the conquest of the country and the overthrow of Saddam Hussein's regime.

31. This figure and the others like it in this chapter are illustrative. In reality, there is no bright line separating the possible combinations of intensity and duration from impossible ones, as the sharp borders here might seem to imply; instead, the spaces demarcate zones in which wars driven by the mechanism in question typically will fall.

32. See also Reiter (2009) for this argument.

33. For the riskiness of the Schlieffen Plan, see for example Ritter (1958). For the riskiness of unrestricted submarine warfare, see Goemans (2000, 95–98).

34. Reiter (1995).

35. Reiter (2009, 102). In a similar vein, albeit a different context, Fearon (2004) argues that "sons of the soil" wars are particularly difficult to resolve because the rebels believe that the government is beholden to a constituency that will continue to push for effective expropriation of land belonging to members of the rebels' ethnic group.

36. Reiter (2009, esp. ch. 3).

37. Schroeder (1994, e.g. 496) describes Napoleon Bonaparte in such a manner, albeit on the basis of limited direct evidence.

38. The paragon of the committed aggressor is Adolf Hitler—see, for example, Schweller (1998)—yet even he was open to political compromises for the first six years of his rule, and as is discussed in greater detail in chapter 4, would have preferred a militarily reasonable political settlement with Britain in the summer of 1940 to continued war.

39. Quoted in Armstrong (1961, 18). Elsewhere Roosevelt averred that changing the noxious German philosophy might require two generations of military occupation.

Foreign Relations of the United States (hereafter *FRUS*) (Washington, DC: Government Printing Office, 1944), vol. I, 501–502.

40. Peterson (1932, 10–11); *New York Times*, "Khomeini Dismisses Truce Offer, Vowing a Fight to the End," October 1, 1980, A1.

41. Blainey (1988, 293).

42. In theory, a leader faced with an opponent who is believed to be implacably committed to aggression might still be open to a peace deal, despite the understanding that it will necessarily be temporary, if she believes that changes in the intervening time will permit her to wage war more effectively once it resumes. In practice, however, in this situation leaders typically worry that peace will restore the initiative to the aggressive opponent, who can choose when and how to resume the war, while the public will relax under the belief that the war is over and hence will not be prepared for the war's resumption. In these situations, leaders also occasionally note that a deal that permits them to prepare for the next round of fighting, for example by rearming, will presumably be rejected by the opponent for precisely that reason.

43. *FRUS*, 1944, vol. 1, 502.

44. For the Russo-Hungarian conflict, see Molnár (1971) and Györkei and Horváth (1999). This situation also describes one of the better-known examples of refusal to negotiate within war from history, namely the Roman decision to utterly destroy Carthage in the Third Punic War. In practice, while the first two Punic Wars were epic clashes, by the time of the third Carthage had been reduced to a Roman tributary state that had no prospect of effectively defending itself. Recognizing this situation, the Carthaginians surrendered hostages and all their weapons and armor rather than fight, but balked when the Romans demanded that they abandon the city and permit it to be razed. The war itself consisted of an extended siege followed by a single battle, with an outcome entirely consistent with expectations. Lloyd (1977, ch. 16), Caven (1980, 273).

45. More precisely, this policy dramatically restricts the range of acceptable settlements to those that address the dispositional commitment problem, minimally through regime change in the country in question. Thus, for example, the American leadership ultimately decided to accept Japanese surrender in World War II because the terms of the surrender would permit them to undertake the fundamental reform of the Japanese political system that they deemed necessary. Hasegawa (2005).

46. The Iran-Iraq War constitutes a marginal case, as it did ultimately end in a negotiated settlement, but only after eight years during which the Iranians refused to consider negotiations with Saddam Hussein's government; it is this extended refusal that justifies the case's inclusion in this discussion. Separately, the two World War II cases obviously overlapped, and thus some readers may be reluctant to consider them to be separate examples. As is apparent in chapters 4 and 5, however, a closer look at the history reveals that the same mechanism operated basically independently but roughly simultaneously in the two conflicts.

47. On Roosevelt's fears, see Dallek (1979, 175). For a discussion of the evidence behind such claims, see chapter 4.

48. Washburn (1871, vol. 2, 185–186, 195–198, 203–205).

49. Some readers may object that principal-agent conflicts logically also should produce wars to the death, by the same basic dynamic described here. I disagree for two reasons. The first concerns the internal checks placed on leaders: the requirement that the public be kept on board places a limit on what leaders can reasonably claim

to be militarily possible and hence on what sorts of war aims the initiators of these conflicts will pursue, in turn making it less likely that opponents will see the initiators as dispositionally committed to war. Second, given the deception at the center of the initiating leader's strategy in a principal-agent conflict, the public in the initiating country typically will prefer to replace that leader rather than fight a defensive war to the death on her behalf. Thus, for example, to the extent that the Tanzanians concluded that Idi Amin posed a fundamental obstacle to a viable peace in the Uganda-Tanzania War, they were aided by the unwillingness of the Ugandans to die to keep him in power.

50. This argument has the additional implication that leaders who are particularly predisposed to explain behavior in dispositional terms—as for example may have been the case with the theocratic government in Iran during the Iran-Iraq War—will be more likely to conclude that their opponent is a war lover who will continue to launch wars until removed from power. Technically, these actors would have a prior belief function that allows for a higher baseline probability that any given actor is dispositionally aggressive; given this higher prior probability, less evidence of a disposition for aggression would be necessary to tip the actor over to believing that it faces a dispositional commitment problem. Developing an *ex ante* testable hypothesis from this point is difficult, however, given the difficulty of observing prior beliefs. One possibility would be that actors that have less experience with international politics would be more likely to have these sorts of prior beliefs.

51. For the security dilemma, see Jervis (1978).

52. Leffler (1992, 49–54, 203–206). This case has of course been minutely dissected, with different schools allocating blame for the Cold War in quite different ways. That American and Soviet officials had different interpretations of the implications of each side's actions is not in dispute, however.

53. Given the significance of misunderstood intentions to this argument, one might wonder why the two sides cannot use costly signals to establish trust. There are several reasons why such an approach is both unlikely to be adopted and, if adopted, unlikely to work. First, under this argument both sides see the opponent as the primary obstacle to peace, and hence would likely believe that the other side should take the first step in signaling benign intentions. Second, as Kydd (2005) notes, while costly signals can produce trust, in situations of significant distrust signals must be unusually costly (as with Gorbachev allowing Eastern Europe to leave the Warsaw Pact) to be credible. The initiator of the preventive war will be reluctant to send such signals, as doing so will likely entail abandoning the initial goal of the war and hence giving the opponent the long-term advantage; the target of the preventive war will be disinclined to send such signals both because doing so will hamper efforts to enforce unconditional surrender and because a truly dispositionally aggressive opponent would simply swallow the concessions and continue the attack.

54. Sincere demands for unconditional surrender are conceptually distinct from, if sometimes related to, the more common phenomenon of foreign-imposed regime change. Most cases of foreign-imposed regime change involve great powers intervening in the domestic politics of minor-power neighbors, with limited or even no fighting in most cases. Given these differences, it would be theoretically inappropriate to test predictions about unconditional surrender on cases of foreign-imposed regime change.

55. Blainey (1988), Fearon (1995), Van Evera (1998), Johnson (2004).

56. See for example Smith (1998a), Gartzke (1999), Filson and Werner (2002), Slantchev (2003b), Powell (1999, 2004), and Smith and Stam (2004).

57. Fearon (1995) points out that divergent expectations alone are insufficient for war; it also has to be the case that actors are unable to share the bases for their differing expectations credibly. Given conflicting preferences, however, actors have an incentive to claim to be strong or resolved even when they are not, because successful bluffs will lead to a better political settlement. This incentive undermines credible signaling, however, because weak or irresolute actors will claim to be more formidable than they actually are, leaving observers uncertain whether claims to be willing to fight are sincere or bluffs.

58. Fey and Ramsay (2007), building on Aumann's (1976) observation that rational actors cannot agree to disagree, have argued that arguments about mutual optimism provide a theoretically incoherent explanation for war. Slantchev and Tarar (2011) however convincingly argue that this result hinges on strong assumptions about exogenous settlement terms and especially the ability of either party in a dispute to impose peace on the other.

59. This observation thus explains why a popular "puzzle"—why a weaker power would ever go to war against a stronger one—is in fact less puzzling than it at first seems. Typically, scholars who advance this puzzle assume that the weaker power is starting a war that it should know that it cannot win; as this discussion demonstrates, however, it is quite possible that, even if it cannot win militarily, the weaker power might win politically, which is after all what matters.

60. Bensahel et al. (2008), Woods et al. (2006, 30).

61. May (2000, ch. 18).

62. Oren (2002, e.g. 151, 172).

63. Jarausch (1969).

64. Hjeholt (1965, 1966), Bucholz (2001).

65. Mack (1975), Mueller (1980).

66. al Marashi (2003), Wrede-Braden (2007).

67. There is some debate in the literature about whether private information is a necessary condition for divergent expectations; from a practical perspective, one could easily imagine that leaders might reach divergent expectation even on the basis of the same information, whether because of psychological biases or simply because of the sheer complexity of international politics. Kirshner (2000), Johnson (2004). Ultimately, however, this distinction is not one that is of great significance for this project. From a practical perspective, distinguishing between divergent interpretations of common information and divergent expectations based on private information is impossible, as in every case some information is available to both sides and some is private, while determining the precise bases on which leaders form their expectations is notoriously difficult. For my purposes, the more important question is less where divergent expectations come from than what happens once they produce a war.

68. Blainey (1988, 56). For formal demonstrations of this process, see Wagner (2000), Filson and Werner (2002), Slantchev (2003b), and Powell (2004). Note that saying that beliefs converge is not the same as saying either that the two sides reach complete agreement or that both sides know who would ultimately triumph militarily. All that is necessary is that the participants roughly agree on the *relative probability*

that each side will win; they can then construct a settlement that gives each side something at least as good as its certainty equivalent to the lottery associated with continued fighting.

69. Wittman (1979).

70. For the details of this war, which is also referred to as the Second Austro-Sardinian War, see King (1967) and Blumberg (1990). It is worth noting that this case is a war with major powers on both sides in which fighting ended in short order; the short duration is likely the reason why many scholars (incorrectly) omit it from the list of major power wars in the past two centuries.

71. Coox (1985, 578, 921).

72. Reiter (2009, esp. 122) concludes that this case—the most limited war that he examines—is particularly well captured by informational dynamics. See also Van Dyke (1997) and Edwards (2006).

73. Powell (2004), Smith and Stam (2004). See also Iklé (1991) for the argument that from a rational perspective wars driven by divergent expectations should not last long.

74. Blainey (1988, 41).

75. For a representative example of explanations for World War I grounded in over-optimism, see Johnson (2004, ch. 3).

76. Gartzke (1999).

77. Reed (2003), Slantchev (2004).

78. Davies (1972).

79. For an early version of this argument, see Kant ([1795] 1957).

80. Simmel (1898), Coser (1956), Levy (1989).

81. Lenin (1920, esp. chs. 5–7).

82. Berghahn (1976), Snyder (1991).

83. Trask (1981, 56).

84. Anderson (1981, 111).

85. For a review of the economics literature on principal-agent problems, see Eisenhardt (1989). There are, of course, significant differences between employer-employee relations and leader-public relations—for one, in the latter the principal (the public) is far more disaggregated, raising potential collective action problems—but the basic dilemmas that a principal-agent dynamic raises are still relevant.

86. Bueno de Mesquita et al. (2003).

87. Goemans (2000).

88. Ibid.

89. Similar arguments that partial democracies are particularly prone to engaging in undesirable actions in the security arena include Snyder (1991) and Mansfield and Snyder (2005).

90. Decalo (1989, 111–113), Kasozi et al. (1994, 124–127). This possibility also can prevent rejectionists from extending a war when settlement is on the table, as for example happened when the Kiel mutiny at the end of World War I torpedoed attempts by hardliners in the German navy to launch a fresh offensive with the goal of undermining settlement talks. Iklé (1991, 70).

91. For the French mutinies, see Pedroncini (1967) and Rolland (2005).

92. Stanley (2009), Stanley and Sawyer (2009), Croco (2011).

93. Hasegawa (2005, 213–214).

94. Khadduri and Ghareeb (1997, chs. 11–12).

95. There exist two standard logics of diversionary war. One posits that leaders can use war to exploit in-group/out-group effects, rallying a divided society against an external threat, at least so long as internal divisions are not too severe. Simmel (1898, 1955), Coser (1956). An alternate view sees war as a noisy signal of leadership competence: good leaders who have been unlucky in domestic politics or incompetent leaders who hope to get lucky in foreign policy may have an incentive to start a conflict so as to improve public perceptions of them and thus increase the probability that they are able to remain in office. Richards et al. (1993), Smith (1998b), Tarar (2006).

96. For typical examples of studies that found a relationship between domestic political conditions and the use of force, see Ostrom and Job (1986), Nincic (1990), James and Oneal (1991), Gaubatz (1991), and Wang (1996).

97. Gaubatz (1991).

98. Chiozza and Goemans (2003). Relatedly, Moore and Lanoue (2003) note that the diversionary hypothesis seems in tension with the observation that domestic economic difficulties are a far more robust predictor of the use of force than presidential approval ratings, given that the president should only need to divert attention from the economy if its troubles are affecting his popularity.

99. For evidence of significant popularity gains in external crises in the United States and elsewhere, see Mueller (1970, 1973), Sprecher and DeRouen (2002), and Lai and Reiter (2005). For the finding that the gains from the average crisis are small, see Lian and Oneal (1993), James and Rioux (1998), and Baker and Oneal (2001). For the finding that the big gains occur in precisely those crises that are least open to manipulation, see Chapman and Reiter (2004) and Lai and Reiter (2005). Colaresi (2007) similarly emphasizes the role of constraints on rally effects.

100. Iklé (1991).

101. For more discussion of this case, see chapters 4 and 5.

102. Mueller (1994, 128–129).

103. See Schlafley (1999) for an example of such speculation, and Hendrickson (2002) for a skeptical view.

104. Indeed, even given the general consensus in World War II, Roosevelt still experienced significant criticism. Prior to the war, Roosevelt's policies of assistance to Britain and the Soviet Union were controversial, and even after enemy attack had rallied the public to the war, congressional Republicans closely monitored government competence in its prosecution while at times arguing that Roosevelt and Secretary of State Cordell Hull were responsible for the surprise at Pearl Harbor. Darilek (1976, esp. ch. 2), Casey (2001).

105. The agent's private information is an essential component of the most compelling models of diversionary war and gambling for resurrection. As such, it features in all formal models; de Figueiredo and Weingast (1999) provide a particularly clear discussion of its importance. Note that just because leaders have access to better information does not mean that they will necessarily make good use of it. Jervis

(1976, 2006). That said, what the leader does with information available to her is less important than what claims she can make on the basis of its existence.

106. Mamdani (1983, 105–107), Decalo (1989, 111–113), Kasozi et al. (1994, 124–127), Mambo and Schofield (2007).

107. Ellsberg (1971), Berman (1982), Bator (2008).

108. Schuessler (2009).

109. This process could be seen, for example, in the spread of control over India. See Lawson (1993, e.g. ch. 4).

110. That said, even well prior to the development of modern communication technology a significant amount of information about ongoing wars was frequently available, if at a greater delay. Thus, for example, English peasants appear to have been surprisingly well informed about developments in France during the Hundred Years War despite the presence of many obstacles to communication that do not exist today. Updates about events in France were passed back through the churches, where priests reported them in sermons; regular citizens seem to have followed developments closely. Seward (1978, e.g. 82).

111. See for example Genova and Greenberg (1979) and Krosnick (1990).

112. Gartner (2008). Likewise, there is evidence that draft-eligible Americans during Vietnam were more likely to favor an immediate withdrawal when their probability of being drafted increased. Bergan (2009).

113. Gartner (1997).

114. This perspective is consistent with the selectorate theory of international politics, which emphasizes the importance of the size of the group of people who influence the selection of the leader. Bueno de Mesquita et al. (2003). Leaders with small selectorates—generally autocrats—face fewer constraints in exploiting members of the general public who do not influence leadership selection.

115. For the argument that democracies tend to be more selective in initiating wars, see Reiter and Stam (2002) and Slantchev (2004).

2. Research Strategy and Statistical Tests

1. Indeed, focusing on well-known cases may be particularly perverse if there are systematic differences related to the mechanisms in question that lead some cases to be well known while others are not. Thus, for example, I argue that in policy wars—nondiversionary principal-agent conflicts—leaders will seek to limit the availability of information about the war. This strategy, if effective, may mean that subsequent historians have difficulty determining everything that happened in the war, as indeed appears to have happened in the Franco-Turkish conflict discussed in chapter 7.

2. More specifically, the cases in the statistical dataset were divided according to length—delineated by a relatively natural break between wars shorter than and longer than one year—and intensity, where conflicts were coded as more intense if they were above the seventy-fifth percentile either in absolute war intensity (deaths per unit time) or in war intensity adjusted by population (deaths per capita per unit time). One case was then selected randomly from within each category. This approach admittedly does involve selection on the dependent variable; if possible, it would have been preferable to select randomly from among the list of wars driven by each of the different mechanisms. Compiling such a list, however, would have

required extensive research on every case in the dataset prior to case selection, an infeasible proposition. As the relevant case studies demonstrate, the wars selected do provide significant variation in the independent variable, as we would have hoped.

3. For detailed discussions of the criteria for identifying wars, see Singer and Small (1972) and Sarkees and Wayman (2010).

4. Other studies that disaggregate multilateral wars in this way include Bennett and Stam (1996), Goemans (2000), and Reiter and Stam (2002).

5. Ayache (1981, 143–144), Balfour (2002, ch. 1).

6. This information is available on an IQSS Dataverse, at http://dvn.iq.harvard.edu/dvn/dv/weisiger.

7. Lacina and Gleditsch (2005).

8. Clodfelter (2007). The most common data source for battle death data is the Correlates of War dataset, which focuses on the slightly broader category of battle-related fatalities. See Sarkees and Wayman (2010, 49–52). This approach, however, results in some anomalies, such as the Ottoman Empire suffering twenty thousand war deaths in the Second Balkan War despite fighting no battles (two thousand soldiers did die of disease), or US deaths in the Mexican-American and Spanish-American Wars being higher than those of their opponents, contrary to all historical accounts, because deaths from illness are only available on the American side. In practice, however, the statistical results presented below are substantively identical when substituting the COW death figures.

9. Determining the dates of war onset and termination can be trickier in civil wars, where fighting is more likely to escalate gradually as rebels establish themselves or to gradually taper off into a stalemate.

10. Fazal et al. (2006).

11. For example, in the nineteenth-century Pacific War, Bolivia withdrew militarily from the conflict in 1880 (leaving Peru to oppose Chile unaided) but only signed a peace agreement in 1884, several months after the Peruvians had capitulated; COW codes the Bolivian-Chilean conflict, and hence the war, as continuing even after the Peruvian capitulation.

12. For a discussion of these costs in the context of the Iraq War, see Stiglitz and Bilmes (2008).

13. Capella (2012). I am deeply indebted to Rosella Capella for sharing this data with me.

14. Maddison (2003). Calculating war cost is a highly imprecise science. Given an initial report of total expenditures, I converted any value in US dollars or British pounds into 2012 dollars using standard conversion rates. For expenditure totals in other currencies (e.g., francs or rubles), I first converted the total into same-year dollars either using known historical exchange rates or, where exchange rate data were unavailable, by using the precious metal content of currency to calculate an effective exchange rate. Thus, for example, I converted Chinese expenditures in the First Sino-Soviet War, which were reported in Chinese taels, into dollars by determining that the tael contained roughly thirty-five grams of silver, which at that time had a value of fifty-nine cents.

15. This approach differs from the strategy in the power transition literature of identifying points at which one country passes another, or alternately points at which countries are relatively equal in power. From a theoretical perspective, the power transition approach is unattractive because it downplays the central dynamic identi-

fied by theory, which is fear of significant future decline. Indeed, bargaining model theory indicates that there is no theoretically coherent reason to believe that power shifts should be more dangerous at points of relative equality than at other points along the dimension (see Powell [1999]). Thus, for example, the Bush administration clearly feared an incipient rise in Iraqi capabilities prior to the 2003 war. Although a nuclear-armed Iraq would still have been far weaker than the United States, an increase in Iraqi capabilities would have limited American freedom of action in the Middle East and would have permitted Saddam Hussein to pursue new policies antithetical to American interests.

16. William Watts, "Americans Look at Asia," A Henry Luce Foundation Project, New York, 1999, 39; Ishihara (1991); Friedman and LeBard (1991).

17. This approach admittedly will not capture anticipated shifts related to qualitative technological advances, such as (most obviously) the acquisition of nuclear weapons. The 2003 Iraq War provides the only case in which fears related to a purported nuclear weapons program produced war, however, meaning that this deficiency should not greatly bias statistical results.

18. Singer et al. (1972). The international relations field has never developed a fully satisfactory measure for power, nor, for that matter, is it likely to be able to do so; see Baldwin (1989) for a discussion of the central problems. Basic material capabilities, captured by variables like population, economic production, and military capacity, end up forming the basis for any measure of power that spans a wide range of countries and time, with the National Military Capabilities dataset the overwhelming choice. This dataset is certainly open to criticism, among other things because it treats an inherently relational concept in a fundamentally nonrelational way, because it fails to include important determinants of state capacity such as geography, and because it focuses purely on measurable capabilities, thus ignoring "soft power" and other nonmaterial forms of influence (see Nye [2004]). However, attempts to improve on existing measures, as for example in Organski and Kugler (1980), generally produce operationalizations that differ only marginally from the existing approach. To ensure comparability with the rest of the discipline, therefore, I use the standard dataset.

19. This approach unfortunately does produce a number of cases for which I have missing data, because at least one participant did not exist five or ten years prior to the war. For those cases in which data were available for at least five years prior to the war, I filled in missing data for ten-year shifts by extrapolating backward for cases in which the country existed but did not meet COW's requirements for system membership and by using data from the year of independence for cases in which the country genuinely did not exist ten years prior to the war. Alternate approaches to handling missing data in these cases, including substituting the size of the shift over the five years prior to war or simply excluding the observations, had no substantive effect on the results.

20. The intensity of fighting is, of course, only one possible indicator of the speed with which participants learn new information, but it is also by far the most theoretically and empirically appropriate. Thus, for example, one might argue that information should be revealed by allies failing to come to a country's aid. This scenario is quite possible; it is also possible, however, that the ally's decision to intervene might be equally surprising for the opponent. Moreover, in many cases, as with the Russian decision to aid Serbia or the Italian decision not to assist Germany and Austria-Hungary in World War I, the ally's decision is no surprise to either side in

the conflict. There is thus no clear scenario in which allied behavior is likely to be unusually informative. Another possibility concerns the speed of communication—in the modern era, bad news from Afghanistan reaches the United States almost instantaneously, whereas for much of the nineteenth century communication between Central Asia and the West might have taken months. In practice, however, this situation simply meant that leaders in earlier times were forced to delegate greater authority to local representatives. Thus, in the Anglo-Iranian War discussed in chapter 6, the British delegated the decision to declare war to the government in India, while the Iranians appointed a representative with significant leeway to negotiate in Constantinople and Paris. Communication difficulties nonetheless lengthened the war marginally, if only because it took a month for news of the peace agreement to reach the battlefield, but not by enough for the difference to be statistically detectable, especially given the relatively small sample of interstate wars.

21. Marshall et al. (2010).

22. WIT distinguishes between military and political losers, as in some cases the military victor may fare less well politically. To give an example, in the Austro-Italian component of the Seven Weeks War, Austria clearly defeated Italy on the battlefield, but the Austrians ended up having to cede Venetia to the Italians as part of the price of peace with Prussia. Most of the cases of variation between military and political victors arise either in draws or in circumstances in which postwar diplomacy (like the Congress of Berlin after the 1877–78 Russo-Turkish War) reallocated benefits; as the latter developments occur after the fighting ends, it is preferable to use the military outcome here.

23. Specifically, the only noticeable variations in governance on the losing side occur in World War II (partially democratic Japan contrasted with Nazi Germany and fascist Italy), the Palestine War (monarchical Jordan contrasted with partially democratic Egypt and Syria), and Vietnam (democratic United States contrasted with relatively undemocratic South Vietnam). Alternate approaches to handling these cases do not affect empirical results.

24. Reiter and Stam (2002), Slantchev (2004), Filson and Werner (2004).

25. This threshold is conventional and does not greatly influence results. A variable that simply uses the Polity score of the initiator is similarly associated with shorter wars, but is typically statistically insignificant.

26. Slantchev (2004). For the rare cases in which my dataset includes a war that was not in his, I use terrain codings from a conflict in a geographically similar area, so that, for example, the Austro-Italian component of the Seven Weeks War is given the same terrain coding as the Austro-Sardinian War of 1848.

27. Stinnett et al. (2002). Substituting the basic contiguity score, which includes intermediate categories for countries separated by limited stretches of water, has no effect on the results.

28. Cunningham (2006).

29. Bennett and Stam (1996), Slantchev (2004).

30. I also experimented with using capabilities data from all participants—i.e., including the (generally quite weak) minor participants in multilateral wars—and with discounting capabilities over distance in the manner introduced by Bueno de Mesquita (1981, 105). Neither alteration had any effect on results.

31. Levy (1983).

32. On cultural difference and war, see Huntington (1993, 1996), although Russett et al. (2000), Henderson and Tucker (2001), Chiozza (2002), and Tusicisny (2004) offer empirical critiques. For ideological difference, see Haas (2005, esp. 17, 30) and Owen (2010).

33. Henderson and Tucker (2001). Huntington is not always entirely clear into which civilization he would classify given countries. That said, most uncertainties either concern countries that have not fought wars or do not affect the coding (e.g. whether Israel is coded as a member of the Western civilization or comprises its own civilization is irrelevant, given that it only fights wars against countries in the Islamic civilization). An alternate coding that identifies a civilizational conflict in multilateral wars when at least one dyad crosses civilizational lines differs only in the coding for the Second Balkan War (based on the Ottoman Empire's involvement); results using a modified variable are thus essentially identical to those presented below.

34. Liberal nationalist ideas often directly threatened members of the old order in the nineteenth century, even when the proponents of liberal positions were not particularly democratic. Thus, for example, Italian nationalism (closely allied at the time with liberal forces), which was ultimately co-opted by (monarchical) Sardinia-Piedmont, posed a significant threat to (monarchical) Austria. Similarly, Bartolomé Mitre's liberal leanings alienated him from Francisco Solano López in the Paraguayan War, even though neither side showed any great commitment to democracy (see Leuchars [2002, 17]). (Mitre was admittedly an elected official during the war, but he demonstrated his low commitment to democratic norms by launching two rebellions once out of power.) The division between liberals and conservatives was pertinent in Europe and throughout Latin America, although the exact meaning of the terms differed somewhat across continents.

35. Once again, changing these variables to code a clash if any dyad within a multilateral war crosses ideological lines results in very few changes and thus does not affect the statistical or substantive significance of the key variables.

36. Bennett and Stam (1996). See however Biddle (2004) for a critique of this conceptualization of military strategy.

37. See for example Smith (1998a), Wagner (2000), Filson and Werner (2002), Smith and Stam (2004), Powell (2004), and Powell (2006). For a qualitative perspective that highlights the same two forms of war termination, see Iklé (1991, 37).

38. A few prior studies of interstate wars have distinguished among types of war termination but along markedly different lines. See Wright (1970) and Pillar (1983). More recent work that has differentiated among types of war termination in a statistical context has focused on victory or defeat (from the perspective of a democratic participant or of the initiator). Bennett and Stam (1998), Slantchev (2004), DeRouen and Sobek (2004). Scholars studying civil wars have generally shown a greater interest in the ways in which wars end, although here too some differences with the simple conquest/settlement dichotomy that follows from theory exist. See for example Licklider (1995), Mason and Fett (1996), and Walter (1997, 2002).

39. The possibility of guerrilla resistance after conventional conquest at times complicates this distinction. For a discussion of this issue, see Weisiger (2012).

40. The online appendices contain the verbatim coding rules, the coding decisions, and explanations for decisions in all potentially questionable cases.

41. For a summary of duration analysis, see Box-Steffensmeier and Jones (2004).

42. I use the semiparametric Cox specification instead of more parametric specifications like the popular Weibull because it imposes fewer assumptions about the

shape of the baseline hazard rate (roughly the probability that a war would end on any given day, which may vary substantially over the course of the war). The Cox specification does assume that the effect of variables on the hazard rate is proportional over time (so that if a variable reduces the baseline probability of settlement by half a month into the war, it will have the same effect several years into the war). Tests of the proportional hazard assumption consistently produce no evidence of violations, either for the model as a whole or for specific variables. I also conduct robustness checks with a log-normal specification—which both the Akaike Information Criterion and comparison of log-likelihoods indicate is the appropriate parametric specification—with results that are consistently substantively unchanged.

43. Fine and Gray (1999).

44. There are thousands of ways in which one might combine the different control variables. It is thus unsurprising that some combinations result in the power shift variable losing significance, although even then it typically is close to significance. I have been unable to find a specification for any dependent variable in which the sign is reversed.

45. I do not include a similar graph for government spending, again for reasons of space. Results are analogous, however: working from model 6 and holding other variables at their median, predicted total spending as a share of GDP is over four times higher when the capability shift variable is at its ninetieth percentile than when it is at its tenth percentile.

46. Goemans (2000).

47. An alternate approach would be to code a major power war as occurring whenever a major power is involved, rather than only when there is a major power on each side. Measured in this way, major power conflicts are no longer than nonmajor power conflicts, although they do continue to be markedly deadlier. Other variables in the relevant regressions are unchanged.

48. I present results only for cultural difference, given overlap with ideological difference. Either variant of the ideological difference variable is, however, consistently statistically insignificant.

49. Fine and Gray (1999).

50. To save space, I present only models that include all explanatory and control variables. Results are consistent in sparser models, however—for example, results for all variables are consistent in every regression if the war intensity variable is dropped from the regression.

51. The findings for duration until settlement are particularly robust here—whereas in the pooled analysis (as in table 2.2), the shifting power variable is frequently insignificant in the absence of statistical controls, in the competing risks specification the variable is always a significant predictor of duration until settlement and is typically a significant predictor (in the opposite direction) of duration until conquest.

52. As before, introducing a quadratic term reveals no evidence that losing partial democracies are particularly averse to settlement.

53. An alternate testing strategy would be to bifurcate the sample into relatively short and relatively long wars and run the analysis separately on each, with the expectation that increased war intensity would be associated with quicker settlement in short wars but with slower settlement in long wars. Conducting these tests, which are reproduced in the online appendix, produces precisely the predicted results; as

this approach raises concerns about selection on the dependent variable, however, I focus on the full-sample tests here.

54. See for example Frieser (2005) for a discussion of the riskiness of Germany's use of blitzkrieg—a maneuver strategy—in World War II.

55. Reiter and Meek (1999). Their study looked not at wars but at the peacetime plans of a set of countries in particular years, with specific country-years chosen at random.

56. To facilitate interpretation, reported results are for total production divided by 1,000,000. This proxy is conventional given that reliable GDP data for a wide range of countries extend back only to World War II. Reiter and Meek also use each side's energy consumption—also a component variable in the NMC dataset—in a robustness check; substituting that variable produces similar results. Substituting interpolated GDP estimates from Maddison (2003) results in a substantial reduction in the total number of observations, but otherwise produces analogous results.

57. To avoid overstating the degree of learning, I code this variable as 0 when a single actor in a multilateral war fights a series of campaigns using a maneuver strategy, unless of course that actor had relevant experience prior to the war. This approach avoids, for example, attributing the German use of blitzkrieg against France in World War II to the success of a similar strategy against the Netherlands.

58. The one exception concerns iron/steel production, which is negative in both reported regressions, although it is positive but insignificant in many robustness checks. This result holds substituting energy consumption for iron/steel production as the measure of economic development.

59. Further analysis, contained in the online appendix, demonstrates that these results are consistently robust. The sole exception arises in a robustness check of the strategy analysis that reaggregates the disaggregated multilateral wars—the lack of robustness to this case is unsurprising given that more than half the cases in which a participant is coded as using maneuver occur during World War II.

3. War to the Death in Paraguay

1. Given the severity of the Paraguayan War, it is frankly astonishing that it has not garnered significant attention from political scientists. To my knowledge, only Abente (1987) and Schweller (2006) have seriously applied political science concepts to explain the war, and neither attempts to explain the failure to reach a settlement once the war was underway. Initial historical literature, written mostly either by Argentines or Brazilians or by participants who were greatly influenced by Paraguayan leader Francisco Solano López's despotism in the latter stages of the war (e.g., Thompson 1869, Washburn 1871), pinned the blame for the war squarely on the Paraguayan dictator's supposed unjustified aggression (see also Box 1929). As time passed, however, there was some recognition of the difficult situation that he faced (e.g., Phelps 1975). Coinciding with revisionist histories of the Cold War, dependency theorists produced a wave of studies whose ultimate villain was economically liberal Britain, which purportedly financed the Triple Alliance to eliminate the nascent Paraguayan system of state socialism as an alternative to capitalism. Fornos Peñalba (1982) provides the best English-language summary of this perspective; Bethell (1996) briefly but thoroughly debunks it. See also McLynn (1979) and Abente (1987). The most recent historical literature, especially Whigham (2002) and Leuchars

(2002), has, without downplaying López's significant failings, explored in greater detail the challenges that he faced, which ultimately provide the basis for a significant reinterpretation of the history of the war.

2. Significant debate has surrounded the number of Paraguayans who died in the war, as the general limits of population records of the time and the destruction and dislocation associated with the war make precise estimates impossible. For an early estimate that deaths were around two-thirds of the total population, see Box (1929, 179). More recently, two studies have reexamined the issue, with one concluding that initial estimates were far too high while the other concluded that they were roughly accurate. See Reber (1988) and Whigham and Potthast (1999), as well as critiques of the two efforts by Whigham and Potthast (1990), Reber (2002), and Kleinpenning (2002). The war severity data used in the statistical analysis here focuses on battle deaths and thus omits the huge numbers of Paraguayans who died of starvation or disease (and who are harder to count), but even so the war is eclipsed in per capita deadliness only by the World Wars.

3. Box (1929, ch. 7), Stewart and Peterson (1942, 178–183), Saeger (2007, esp. 10–12).

4. The declaration came in 1813; Brazil was the first country to recognize Paraguayan independence, which it did (largely to gain Paraguayan support against Argentina's Juan Manuel de Rosas) in 1844. Leuchars (2002, 2, 23).

5. For the possibility of that these regions would gain independence, see for example Box (1929, 276), Whigham (1991, 53–56), and Rector (2009, ch. 5).

6. Given that the focus of this study is on political decision making, the summary of the military side of the war will be quite brief. For more details on the military campaigns, see Leuchars (2002).

7. The most detailed account of Paraguayan capabilities prior to the war is Whigham (2002, ch. 7).

8. For an overview of work on the Paraguayan state's role in the economy, see Pastore (1994).

9. Whigham (2002, 217).

10. Leuchars (2002, 56).

11. The overthrow was the culmination of the La Plata War of 1851; see Lynch (1981) and Whigham (2002, 119–121).

12. Quoted in Leuchars (2002, 39). See also Saeger (2007, 82, 93) for evidence of close relations between López and Urquiza.

13. Whigham (2002, 220).

14. Historians have attributed Urquiza's loyalty to a range of considerations, including Mitre's solicitousness, the recognition that Argentina shorn of Buenos Aires was not a viable country, concerns about López's plans and war-weariness, a possible desire to succeed Mitre in the next elections, and putative personal financial gains from the alliance with Brazil. McLynn (1979), Katra (1996, 257–258), Whigham (2002, 221).

15. de la Fuente (2004).

16. Whigham (2002, 418).

17. It is also worth noting that Mitre in Argentina and Emperor Pedro in Brazil overestimated the ease with which Paraguay could be defeated. Whigham (2002, 217, 272–273). This overestimation likely contributed to the refusal to attempt to allay

López's concerns, as neither side seems to have believed that the Paraguayans would actually start a war.

18. For evidence of prior Paraguayan attempts to reach a final understanding on borders with both Brazil and Argentina, see Whigham (2002, chs. 4–5).

19. Leuchars (2002). Indeed, Buenos Aires's ability to restrict Asunción's trade under the old Spanish Viceroyalty had been a major reason why the Paraguayans desired independence upon the collapse of the Spanish Empire. Humphreys (1957, 619).

20. For the history for the border disputes, see Box (1929, chs. 2–3) and Williams (1979, chs. 9–10).

21. Whigham (2002, 78, 85–92).

22. Whigham (2002, 213). See also Box (1929, 278).

23. For Paraguayan concerns about these shipments, see Whigham (2002, 213–215). In practice, many of the cannons that the Brazilians shipped to Mato Grosso were ancient and worthless, and had been included in shipments only as ballast. (Personal communication with Thomas Whigham.)

24. Attempts to reunify old Spanish administrative units that had broken apart upon independence frequently underlay wars in Latin America in the nineteenth century. To cite some of the main examples, Chile went to war to break up a Peruvian-Bolivian Confederation, Colombia fought a brief war against Ecuador with the goal of re-creating Gran Colombia, and the Guatemalan Justo Rufino Barrios made several attempts to forcibly re-create the Central American Union (see Scheina [2003]). Had it managed to deal effectively with its internal divisions, Argentina easily might have tried to force Paraguay and Uruguay into union.

25. Whigham (1991, 53–56).

26. See McLynn (1979) for an analysis that highlights the centrality of changes in Argentine domestic politics to the decisions that led to the war, although his interpretation grossly overstates the degree to which Mitre intended to bring about the war that his appearance produced.

27. Fazal (2007).

28. Quoted in Katra (1996, 257).

29. MacLean (1995).

30. Plá (1976).

31. Saeger (2007, 104). The seriousness of this potential threat can be seen in a previous crisis in the 1840s, in which an Argentine river blockade—to which "war was the only reply"—helped provoke the Paraguayans into a declaration of war. In that case, the crisis ended without a direct clash, with the Argentines dropping the blockade. Box (1929, 20–22).

32. Box (1929, chs. 2–3).

33. Leuchars (2002, 29).

34. Thus, for example, López's protest was met "with shouts of laughter" and with recommendations to its author "to attend to the state of his huts and settle the squabbles of his half-naked squaws at home." Quoted in Whigham (2002, 158).

35. The terms of the Treaty of Triple Alliance obviously could not influence López's decision for war, as the treaty had not yet been signed when López decided on war, and its territorial clauses remained secret until the British published them in early 1866. Whigham (2002, 279–280). That said, the nature of the territorial terms provide

[236]

good reason to consider the war indeed to have been "a war over the partition of Paraguay," precisely in line with López's fears. Leuchars (2002, 46).

36. Quoted in Leuchars (2002, 28–29).

37. Thompson (1869, 25).

38. In the 1839–52 Uruguayan civil war, the Blancos, acting with Argentine support, overran rural Uruguay but were unable to take Montevideo. They besieged the city starting in 1843 but were unable to prevent supply from the sea; outside assistance thus permitted the Colorados to hold on until the anti-Rosas coalition under Urquiza invaded in 1852 and forced the Blancos to surrender.

39. See for example Leuchars (2002, 146).

40. Details on the exact content of the discussions are unfortunately unavailable. For a summary of the available evidence, see Cunninghame Graham (1933, 199) and Leuchars (2002, 145–147).

41. See for example Washburn (1871, vol. 2, 203–205).

42. That said, the need to strike into Uruguay, explained by the preventive motivation, was the reason that he was willing to contemplate the risk of attacking Argentina in the first place.

43. For examples, see Leuchars (2002, 145–147). The differing views on whether to negotiate with López was a significant source of friction between the allies; Emperor Pedro at one point complained that Mitre "drags his feet and aims to drag me into a peace which our honor does not let us accept." Quoted in Bernstein (1973, 103). The Uruguayan force was too small to be of significance in this case; its leadership does, however, seem to have been open to talks.

44. Leuchars (2002, 147).

45. Washburn (1871, vol. 2, 185–186, 195–198, 203–205), Peterson (1932, 16–17), Cunninghame Graham (1933, 133), Phelps (1975, 164–165).

46. Bernstein (1973, 106–107).

47. On rebellion in Argentina, see de la Fuente (2004). On the connection between the war and eventual domestic change in Brazil, see Bernstein (1973) and Needell (2006).

48. Quoted in Box (1929, 23). Urquiza, whose interests were more aligned with Paraguay's than they were with those of Buenos Aires, did formally recognize Paraguay as independent after taking power in 1851, but the *porteños* did not consider themselves bound by this act.

49. Quoted in Whigham (2002, 278).

50. Warren (1978).

51. For the negotiation and terms of the treaty, see Whigham (2002, 276–281).

52. See Warren (1978, 116) for the terms of the Brazilian-Paraguayan treaty. In addition to granting the Brazilians the extent of their prewar territorial claims (but no more), the treaty also called for Paraguay to pay a tremendous indemnity, but the Brazilians also made it clear that they would not require payment so long as the Paraguayans cooperated in limiting Argentine gains, something that Paraguay's leaders were more than happy to do.

53. Peterson (1932, 10–11).

54. Leuchars (2002, 46). For other examples of Brazilian intransigence (beyond the repeated refusals to negotiate highlighted above), see Kolinski (1965, 125), Phelps (1975, 167), and Barman (1999, 230).

55. McLynn (1979, 21–22).

56. Whigham (2002, 121).

57. Saeger (2007, 81).

58. Warren (1978).

59. Whigham (2004, 179, 195).

60. Kraay (2004), Bernstein (1973, 107).

61. Barman (1999, 356–361). From a domestic political perspective, it is also worth noting that the war was backed by both liberals and conservatives in Brazil. Bernstein (1973, 91–92).

4. World War II

1. Charmley (1993, esp. 647–649).

2. Hildebrand (1970), Hillgruber (1981).

3. Taylor (1961), Broszat (1966, esp. chs. 3–4), Mommsen (1991, esp. chs. 7–8).

4. For an essay that helped clarify the terms of the functionalist-intentionalist debate, see Mason (1981). Browning (1992) provides a relatively recent influential functionalist perspective, while Goldhagen (1996) advances an extreme (and historically unconvincing) intentionalist perspective.

5. Thus, for example, Broszat (1966, esp. 51) argues that Nazi ideology was fundamentally incoherent (and hence that it cannot be seen as the basis for the Holocaust), but he acknowledges that race politics and the acquisition of Lebensraum—the key elements of ideology for my argument—were present early and basically unchanged throughout Hitler's political career.

6. This stance is similar to that of Rich (1973), although I highlight different evidence.

7. His optimism with respect to Britain was repeatedly in evidence in *Mein Kampf* and in his unpublished second book, although by the late 1930s evidence of unexpected hostility reduced his confidence. Hitler (1925, e.g. 664–665), Hitler (1928, esp. ch. XIV). On confidence on the eve of war, see Taylor (1952, 266–267) and Blainey (1988, 48–49); Powell (2006, 195–199) provides a contrasting view. Note also that Hitler's optimism prior to the invasion of France was not shared by his generals. May (2000, ch. 18).

8. On preinvasion German confidence, see Cecil (1975, ch. 8). On the purge of the Red Army and related weaknesses, see Gorodetsky (1999, 115) and Reese (1989). For examples of similar expectations of a quick Soviet collapse in Britain and the United States, see Dallek (1979, 278) and Gilbert (2000, 831).

9. Hitler (1925, 653), Hitler (1928, 100); *Documents on German Foreign Policy*, series D, vol. 1, doc. 19 (Washington, DC: U.S. Government Printing Office, 1949), 34.

10. Quoted in Kershaw (1998, 588).

11. See for example Rich (1973, vol. 1, 208–210).

12. See Gorodetsky (1999, e.g. ch. 10) for a discussion of Soviet conciliation of the Germans during the period prior to Barbarossa.

13. For an example of the relatively early recognition among the generals that defeat was coming, see Bullock (1953, 716–717).

14. Cecil (1975, 23). Infighting and the selective use of information within the Nazi hierarchy also limited Hitler's effective policy freedom to a degree that was not recognized at the time. Kershaw (1985, ch. 4), Shore (2003).

15. Press (2004, 151).

16. May (2000, 106).

17. Quoted in Rothfels (1961, 82).

18. Quoted in Kershaw (1987, 145).

19. Lüdtke (1992).

20. Kershaw (1987, 143–147).

21. Copeland (2000, ch. 5).

22. Quoted in Copeland (2000, 131).

23. Bullock (1953, 735–740).

24. Many studies that try to present Hitler as a calculating actor responding to developments in the international system, as this one does, as a first step bracket Hitler's racial ideology, arguing that anti-Semitism and the focus on the German race and German racial purity were important primarily in internal politics. See for example Taylor (1961) and Copeland (2000). This approach typically follows from a desire to separate a rationalist explanation for German foreign policy from Nazi domestic policies—especially the Holocaust—that were particularly horrific. As the discussion below indicates, however, I believe this approach to be fundamentally mistaken: without understanding Hitler's beliefs about the world, as grounded in his ideology, it is impossible to understand why he made the foreign policy choices that he did or what his ultimate aims were.

25. The discussion in this section relies primarily on Hitler's *Mein Kampf*, written in 1924 during his prison stay following a failed coup, and his unpublished second book, completed in draft form by 1928 but never published. For a summary of Hitler's ideology that is quite compatible with the one presented here, see Rich (1973, vol. 1, 3–10, 81–82).

26. See Smith (1986) for a discussion of the long-running competition between two ideological justifications for German imperialism, which he calls *Weltpolitik* and Lebensraum, in which Hitler, although drawing on the economic logic of Weltpolitik, gave policy priority to the goals associated with Lebensraum.

27. Hitler (1925, 131).

28. Ibid., 131, 138. See also 642–643 for the first principle that "foreign policy must safeguard the existence on this planet of the race embodied in the state, by creating a healthy, viable natural relation between the nation's population and growth on the one hand and the quantity and quality of its soil on the other hand" (emphasis removed).

29. Ibid., 133–135.

30. Ibid., 139, 644.

31. Ibid., 646.

32. Hitler consistently rejected calls for a return to the borders of 1914. See, for example, Ibid., 651–652, and Hitler (1928, chs. 8–9).

33. Hitler (1925, 661; emphasis removed).

34. Ibid., 140, 143; Hitler (1928, 76–77).

35. Hitler (1925, 654). An interesting question is why Hitler did not see territory gained through conquests in Western Europe as sufficient to guarantee Germany's security. The answer is partly racial—he saw the French and especially the Dutch and Scandinavians as racially superior to the Slavs, and hence intended to some degree to co-opt them rather than displacing or killing them—and partly strategic, in that even the incorporation of the whole of the western conquests would still leave Germany overshadowed in the long run by Soviet Russia.

36. Hitler (1925, 141), Hitler (1928, 70, 191).

37. Hitler (1925, 654), Hitler (1928, 148–152). See also Kershaw (2000, 285).

38. For the argument that Germans outside the Reich were being lost to the race, see for example Hitler (1928, 99–100).

39. Hitler (1925, 140).

40. Ibid., 138. See also 654.

41. Hitler (1928, 89–90).

42. Quoted in Kershaw (1998, 442).

43. *Documents on German Foreign Policy*, series D, vol. 1, doc. 19, 30.

44. Quoted in Kershaw (2000, 88).

45. Thus, for example, most of his advisors believed that the remilitarization of the Rhineland could have been accomplished through negotiation in another year or two, but he was unwilling to wait. Kershaw (1998, 582, 584). Similarly, in the Munich Crisis, he responded to British willingness to accept the transfer of the Sudetenland by raising his demands, although in this case he ultimately backed down after being convinced that further time for rearmament would favor Germany. Overy (1999), Copeland (2000, 133).

46. For similar arguments about Hitler's beliefs in a deteriorating military situation, see Copeland (2000, ch. 5) and Powell (2006).

47. *Documents on German Foreign Policy*, series D, vol. 1, doc. 19, 34.

48. Ibid., series C, vol. 5, doc. 490, p. 854.

49. Roberts (2006, 16).

50. Copeland (2000, 123–145) makes a convincing case that the generals shared Hitler's fears of decline and were quite prepared for a preventive war, although they differed at points on questions of timing.

51. Hitler predicted inevitable French opposition to German expansion in *Mein Kampf*, although he was more sanguine about a possible alliance with Britain. Hitler (1925, 665), Rich (1973, vol. 1, 4–5). By the late 1930s, however, British behavior and reports from his representatives in London had left him with few grounds for expecting British cooperation. See for example then-ambassador and future foreign minister Joachim von Ribbentrop's January 1938 assessment of British intentions toward Germany. *Documents on German Foreign Policy*, series D, vol. 1, doc. 93, 162–168.

52. For the British view, see Ripsman and Levy (2008).

53. For Hitler, see Chief of Council for the Prosecution of Axis Criminality, *Nazi Conspiracy and Aggression*, vol. 7 (Washington, DC: U.S. Government Printing Office, 1946), 801, 812. For Allied views, see Smart (2003, esp. 8–10). The expectation that an arms race would lead to debilitating inflation in Germany provided a further reason for haste. Klein (1959, 78–79).

54. Roberts (2006, 15–16), Keegan (1990, 129).

55. Kershaw (2000, 44, 285–286). For a contemporary lecture by the German ambassador to the Soviet Union that emphasizes the significant, if temporary, impact of the purges on Soviet capabilities, see *Documents on German Foreign Policy*, series D, vol. 1, doc. 610.

56. For Hitler's focus on the importance of avoiding fighting on two fronts, see for example Hitler (1928, 78–79, 185, 187).

57. *Nazi Conspiracy and Aggression*, vol. 3, 575. In June 1940, a German intelligence report noted that the Russians were disturbed by the prospect of a quick German victory in the West, but that "active participation of Russia in the war is entirely out of the question because of military weakness and inner-political instability." Ibid., vol. 6, 984.

58. Quoted in Gorodetsky (1999, 323).

59. Quoted in Clark (1965, 24) and Warlimont (1964, 112).

60. Quoted in Rich (1973, vol. 1, 240).

61. *Nazi Conspiracy and Aggression*, vol. 6, 1000.

62. For the belief that Britain was holding out in the hope of eventual Russian assistance, see Halder (1988, 227). For an example of the interpretation of Barbarossa as primarily driven by the desire to convince the British to negotiate, see Iklé (1991, 53). For the invasion being Hitler's greatest mistake, see Schulman (1948, ch. 10) and Kahn (1978, ch. 24).

63. May (2000, ch. 18, esp. 267–268).

64. Kershaw (2000, 43).

65. Another source of puzzlement has been his cavalier attitude toward the United States, most notably his declaration of war following Pearl Harbor. This point is ancillary to the primary concerns here, but it is worth noting that by December 1941 the United States was already at war with Germany in everything but name—the navy, for example, had been operating for several months under a standing order to shoot any German vessel encountered in the Atlantic on sight. In this context, Hitler could quite reasonably have seen a declaration of war as simply an acknowledgment of reality, as well as an opportunity to improve the sometimes strained relations with Japan and hence possibly bring about an eventual Japanese attack on Russia in the Far East. For an argument along these lines, see Rich (1973, vol. 1, ch. 20).

66. Domarus (1997, vol. 3, 1845–1846). Remarkably, after the victory in the west, Hitler's demands of Britain in some ways *dropped*, as he no longer explicitly insisted on the retrocession of the colonies (although of course he demanded recognition of the new situation with respect to France).

67. Woodward (1971, ch. 25).

68. Quoted in Frieser (2005, 310–311).

69. Allen (2005). See *FRUS* 1944, vol. I, 484–579, for discussions of German peace feelers in the closing stages of the war, when a wide range of individuals and groups sought to arrange some sort of deal, generally involving the overthrow of Hitler and peace in return for some sort of political concessions from the Allies. Also worth noting is the quixotic personal mission of Rudolf Hess, one of the highest-ranking Nazis, who, officially without authorization, flew personally to Britain in May 1941 in a unilateral attempt to negotiate a peace deal, an act that resulted only in his arrest and incarceration.

[241]

70. Reiter (2009, 115–119). For a recent argument criticizing views that Stalin authorized moves toward peace, see Roberts (2006, 165–166).

71. Mastny (1972), Koch (1975). Separately, as part of pushing for peace with the Western powers, Himmler responded to British and American truculence by threatening to reach a separate peace with the Soviets, with the consequence that Germany would fall to communism. *Foreign Relations of the United States* (hereafter *FRUS*), 1944, vol. I, 491. This threat was seen, almost certainly correctly, as baseless bluster.

72. Rich (1973, vol. 2, ch. 8 and 11 and 394–398).

73. Ibid., 330–331.

74. For comments on the wasted deaths of World War I or on the costs of shedding blood for inappropriate goals, see for example Hitler (1925, 651–652) and Hitler (1928, 85–86).

75. Thus in the closing months of the war Hitler repeatedly returned to the miraculous Prussian escape in the Seven Years' War, when the death of Empress Elizabeth precipitated Russia's withdrawal from the war; the death of President Roosevelt fostered a brief period of hope in the last month of the war. Kershaw (2000, 791–792).

76. See for example Schweller (1994).

77. Other views of Hitler's war aims are possible. Thus, for example, many American policymakers concluded that Hitler's political aims changed over time, increasing in response to the pusillanimity of his opponents. Jervis (1976, 223). The argument against unlimited aims advanced here also addresses the logical claims of such an alternate view.

78. Whether there were territorial limits to how far Hitler planned to go is a separate question, of course, from whether the Allies *believed* such limits to exist. I discuss Allied beliefs in the context of unconditional surrender later in this chapter.

79. For reviews of the relevant debate, see Michaelis (1972) and Kershaw (1985, ch. 6). Most globalists argue that Hitler had an intentional plan to conquer the world, typically through a three-step process of peaceful expansion, a European war, and then a war of world conquest. The most prominent exponents of this view in English are Hildebrand (1970), Hauner (1978), Hillgruber (1981), and Weinberg (2005, ch. 1). A smaller group agrees that Germany would have continued to expand indefinitely but argues that that expansion by domestic political imperatives rather than an innate desire for expansion. See for example Broszat (1970). For the continentalist position that Hitler's aims were limited to the acquisition of Lebensraum in Eastern Europe, see for example Trevor-Roper (1960) and Rich (1973, vol. 2, ch. 12).

80. Rauschning (1940). Rauschning broke with the Nazis in 1934 and wrote the book with the intention of convincing Germans to withdraw their support for Hitler's regime; it subsequently attracted significant attention during the war in Britain and the United States.

81. Kershaw (1998, xiv), for example, comments that the book is "now regarded to have so little authenticity that it is best to disregard it altogether."

82. Moltmann (1961, 201). I would argue that a "world power" in Hitler's terms is a nation with territory sufficient to protect itself and to support its population; by those terms, multiple world powers existed, but Germany was not one of them.

83. Koch (1968, 126).

84. Weinberg (1995, ch. 15), Goda (1998).

85. Moltmann (1961, 224).

86. See for example Weinberg (1995, ch. 15).

87. It should, however, be noted that Hitler is less uniformly dismissive of the United States in *Mein Kampf* than is typically believed, at one point for example contrasting its "immense inner strength" with "the weakness of most European colonial powers." Hitler (1925, 139).

88. For the relevant discussion, see Hitler (1928, ch. 9). Purported allusions to a future war depend on Hitler's claim that only a state that follows his recommendations "will be able to stand up to North America" (116); this comment was a rhetorical aside that in no way implies a planned invasion of the Americas.

89. Weinberg (1964).

90. Compton (1967, 247–248). Compton more generally attributes Hitler's disinterest in the United States to a combination of ideology, which saw Americans as racially mixed and hence degenerate, and geopolitics, in which countries outside the European continent were seen as largely irrelevant.

91. See for example Hitler's speech in Domarus (1997, vol. 3, 1845–1846).

92. Jackson (2003, 11).

93. Smart (2003, ch. 3).

94. Quoted in Jackson (2003, 1).

95. Quoted in Reynolds (1990, 329).

96. Bell (1974, 48–52).

97. Reynolds (2001, 234). See Bialer (1980) for a general survey of prewar British fears about the effects of air power.

98. Gilbert (1994, 157), Domarus (1997, vol. 3, 2062–2063).

99. For popular and scholarly accounts of the War Cabinet discussions, see Knight (1977), Reynolds (1985), Costello (1991), Carlton (1991, 1993), Lawlor (1994), Lukacs (1999), and Reiter (2009, 95–108).

100. Gilbert (1994, e.g. 158). A range of interpretations of these events exist, including for example the claim that Halifax sought to capitulate while Churchill aimed to continue the war or the argument that both Churchill and Halifax were in principal open to negotiations, albeit not on the terms that Hitler was willing to offer. See Costello (1991) and Carlton (1993), respectively.

101. *FRUS, Conferences at Washington, 1941–2, and Casablanca,* 1943, 727.

102. This point is muddied in the historical record by Roosevelt's odd and untrue claim that he advanced the demand as an off-the-cuff comment in a press conference, an interpretation belied both by extensive evidence of prior discussions of the demand and by the conviction with which he subsequently defended the policy. See Sherwood (1950, 696) and Dallek (1979, 373–376) for useful discussions of Roosevelt's behavior in this incident.

103. For details on the advisory committee's discussions, see Notter (1949, Part 2) and *FRUS: Conferences at Washington, 1941–2, and Casablanca,* 1943, 506n2.

104. Notter (1949, 85).

105. Ibid., 101, 127.

106. Quoted in Sherwood (1948, 418).

107. Gilbert (2000, 361), quoted in Dallek (1979, 256).

108. Allen (2005), Hindley (1996).

109. Kecskemeti (1958, ch. 5).

110. Reynolds (1985).

111. Gilbert (1994, 247).

112. Ibid., 337–338, 341. See also 119, 155, 163, and 255 for pessimistic or negative comments by British leaders about the Americans during the critical period. See also Woodward (1962, 78–79).

113. Smart (2003, ch. 3).

114. Gilbert (1994, 157–158).

115. Ibid., 313.

116. With the French collapse, Churchill acknowledged that the blockade was "largely ruined," but he continued to hold out hope that bombardment might weaken the German economy, for example objecting later in the summer to Herbert Hoover's plans to provide food aid to the occupied populations on the grounds that "the front lines run through the factories." Quoted in Gilbert (1994, 441) and Doenecke (2000, 95).

117. Roberts (1991, 177), Neillands (2001, 57–58).

118. Overy (2001, 71).

119. On the severity of the U-boat threat, see Churchill (1950, esp. ch. 7). For discussions of the expectations of a German invasion and its likely consequences, see Hinsley (1979, vol. 1, ch. 5) and Bell (1974, 50).

120. Quoted in Costello (1991, 213).

121. Gilbert (1994, 93, 338).

122. Roberts (1991, 224).

123. On motivated bias, see Jervis (1976, esp. ch. 10).

124. Charmley (1993, 560).

125. Butler (1957, xviii).

126. The Special Operations Executive (SOE), tasked by Churchill to work covertly to "set Europe ablaze," provides a similar example of motivated bias: at its creation, British leaders apparently hoped that it might provoke a general uprising that would provide the coup de grâce originally expected from the French army, but in the event it immediately entered into factional battles within the British bureaucracy and ultimately served as at best an auxiliary to the war effort. In the summary of one historian, the SOE emerged "as a desperate attempt to plug the gaping hole in British strategy caused by the collapse of France, and wildly unrealistic hopes were pinned upon it at the very highest levels—as they also were on other 'indirect' methods of defeating the Germans such as economic warfare and strategic bombing"; one of the organization's leaders described it after the fact as "no more than a hopeful improvisation devised in a really desperate situation." Stafford (1980, 2); quoted in Stafford (2006, 49). For a fuller history of the SOE, see Foot (1966) and Stafford (1980).

127. James (1970), Charmley (1993, 434).

128. Roberts (1991, ch. 21), Lawlor (1994, 33).

129. Trachtenberg (2005), Schuessler (2009).

130. Bell (1990, 16), Gilbert (1994, 22, 243, 246).

131. Casey (2001, 74–75).

132. *Department of State Bulletin*, vol. XI, no. 265, 83.

133. Charmley (1989, 210–212), Self (2006, 437–438). Similarly, the broader cabinet met Churchill's announcement that "whatever happens at Dunkirk, we shall fight on" with general acclaim. Gilbert (1994, 182–184), Charmley (1993, 405–406).

134. Burridge (1985, 129, 132, 145), *FRUS*, 1940, vol. I, 81.

135. Bell (1974, ch. 6).

136. Langer and Gleason (1953, 757–758, 939).

137. Darilek (1976, 18, 24).

138. Ibid., ch. 2.

139. Foster (1983, 20).

140. Berinsky (2009, 47–48).

141. See Powell (2006) and in particular Reiter (2009, ch. 6).

142. Roberts (1991, 178).

143. See for example Chase (1955) and Armstrong (1961).

144. Roosevelt (1946, 117), Carroll (1948, 334), Pozdeeva (1994, 356–357).

145. Chase (1995, 262).

146. The exception to this claim is the argument that Roosevelt somehow "gave away" Eastern Europe to the Soviets. Baldwin (1950, 14), Wittmer (1953). In practice, however, unless the Western Allies were prepared either to invade France far earlier than June 1944—which they ultimately decided they were militarily incapable of doing—or make a deal with Germany to fight the Soviets—which no serious commentator in the West advocated, even in retrospect—Eastern Europe simply was not Roosevelt's to lose. Reynolds (2006, 66–67).

147. For examples of contemporary criticism, see Hull (1948, 113), Chase (1955, 113), Balfour (1970, 719), and Villa (1976).

148. Sherwood (1950, 695). In practice, evidence that the demand strengthened German resistance is scanty, especially in comparison with the clear effect of the aerial bombardment of German cities. Chase (1955, 267).

149. Hull (1948, 1570).

150. Reiter (2009) similarly argues that the refusal to negotiate followed from a mistrust of Hitler's character, but he lacks a strong theoretical explanation for the origin of that mistrust.

151. English translations of the book were quite limited prior to World War II, with the first full translation appearing only in 1939, and the book's poor organization tended to put off those who attempted to read it in German. As a result, relatively little was known about it, which aided Hitler's attempts in international diplomacy to downplay his earlier statements. See Ensor (1939) for a fascinating contemporary discussion of the book, and Barnes and Barnes (1980) for a history of the delay in the book's translation.

152. Indeed, in 1939 Stalin had reason to hope that whichever country lost the war would turn communist, as Russia had done in World War I. A purported secret speech to the Politburo defends the pact with Hitler in precisely these terms

(Pleshakov 2005, 43–44); Sluch (2004) however makes a fairly convincing (if not definitive) case that this speech never occurred, although he acknowledges that the main arguments attributed to Stalin were likely close to what the dictator believed.

153. This point should not be taken as implying that there were no policy differences among these men—clearly such differences existed. The central point is simply that their interpretations of German behavior shifted in similar ways and in response to similar observations over time, and that this shift took all of them from the view that one could negotiate with Germany to the view that negotiation was impossible.

154. Gilbert and Gott (1963, 340).

155. Quoted in Roberts (1991, 185).

156. Ibid., 214. His openness to the approach to Mussolini in the May War Cabinet meetings can also be attributed to greater optimism that France might be induced to continue to fight, a hope that he held onto longer than others. Lawlor (1994, 57). Halifax also apparently did not see a significant threat to British morale from the knowledge that negotiations were occurring; here most historians have sided with Churchill in arguing that negotiation would have undermined the popular will to continue resistance, whatever the terms that Germany demanded.

157. Gilbert (1994, 799).

158. Quoted in Self (2006, 431).

159. Dilks (1978).

160. Quoted in Self (2006, 437–438).

161. Quoted in Gilbert (1977, 452).

162. Quoted in James (1970, 225).

163. Ibid., 228–229.

164. Ibid., 318.

165. Quoted in Gilbert (1977, 1001).

166. Quoted in Kimball (1997, 44).

167. Gilbert (1993, 194).

168. Farnham (1997, 62).

169. Dallek (1979, 144, 149).

170. Kimball (1997, 44).

171. Dallek (1979, 175).

172. Sherwood (1950, 667).

173. Notter (1949, 127).

174. *FRUS* 1944, vol. I, 584–585, 587–588.

175. Ibid., 588–589, 592.

176. Armstrong (1961, 42).

177. Kecskemeti (1958, ch. 4).

178. For a discussion of the varied diagnoses of the German problem among Allied leaders, see Armstrong (1961, 21–22, 28).

179. Casey (2001, esp. ch. 5).

180. *FRUS* 1944, vol. I, 501–502.

181. Armstrong (1961, 19–20).

182. *FRUS: Conferences at Cairo and Tehran*, 1943, 511–513.

183. Gilbert (2000, 322).

184. Ibid., 1513. See also 332, 1428, 1487, and 1659. Churchill ultimately acquiesced to the views of his allies; even had he prevailed, it is unlikely that any German government would have agreed to his requirement that Prussia be permanently alienated from Germany until the military situation approached that of the final months of the war.

5. Additional Commitment Problem Cases

1. This summary of the onset of the war is extremely abbreviated. For useful discussions of the prewar negotiations with contrasting views on the reasonableness of various sides' demands, see Rich (1985) and Goldfrank (1994).

2. On the Russian policy of restraint, see Kerner (1937). The Russians recognized that they held predominant influence at the Porte and that in the event of Ottoman collapse the likely result would be a partition that would saddle them with great power rivals on their southern flank.

3. Seaton (1977, 38).

4. Puryear (1935).

5. This point was generally accepted, although there were important differences of opinion as to when the collapse would occur. For examples, see Curtiss (1979, 65–66).

6. Rich (1985, 38).

7. Argyll (1906, 448). Similarly, Foreign Minister Clarendon, in an address to the House of Lords shortly after the British declaration of war, averred that "there is not a man in Russia that does not believe that Constantinople will ultimately belong to Russia. It will be our duty as far as possible to see that that expectation shall be disappointed. Because were it to succeed … it is not too much to anticipate that more than one power would have to undergo the fate of Poland." *Annual Register* (London: 1854, 57).

8. In this context, it is worth remembering that Russia was the sole major continental power to have been untouched by the revolutions of 1848. In particular, the demonstration of Austrian reliance on Russia—the Austrians had been able to put down Hungarian uprisings only with the assistance of several hundred thousand soldiers from Russia—raised fears that Russia might be in a position to unilaterally revise the status quo with respect to Turkey. See for example Taylor (1964, 91–93), who argues that Russia assisted Austria in Hungary precisely to gain a free hand with respect to Constantinople.

9. For British war aims, see Rich (1985, esp. 107–110). Palmerston, who was prime minister for the second half of the war, and Stratford de Redcliffe, the British ambassador to the Porte, had particularly ambitious aims that entailed significant territorial losses by Russia from the Baltic to the Black Sea region.

10. Without the preventive motivation, it is highly unlikely that the war would have happened. That said, divergent expectations also played a significant role in bringing about the conflict. Most notably, Nicholas consistently underestimated the probability that the British would be willing to cooperate with the French in opposing Russian actions in Constantinople; had he known that the British would be willing

to fight (and that the Austrians and Prussians would not back him), he likely would not have adopted a policy as aggressive as he did. See for example Lincoln (1978, 333–341).

11. Some might object to this argument on the grounds that it would have been unreasonable for the Russians to believe that they could achieve the kind of military success necessary to impose unconditional surrender. While it is no doubt possible that this situation played a role in Russian thinking, it is worth remembering that at the time that American policymakers formulated the unconditional surrender policy in World War II the Americans and their British allies were completed excluded from the Continent as a result of unprecedented German victories; without the benefit of hindsight, a policy aiming at the total defeat of Germany was simply audacious at the time.

12. The lack of clarity in Russian war aims arose both because documentation on Russian decision making is limited relative to the availability of information about the other powers (partly because of the concentration of power in the hands of a few men and partly because of the vagaries of subsequent history) and because the Russians generally were on the defensive and thus simply sought to induce the Allies to limit their demands.

13. Rich (1985, 110).

14. For discussions of wartime negotiations, see for example Rich (1985, ch. 9) and Wetzel (1985, chs. 6–7).

15. Curtiss (1979, 60–61).

16. Schroeder (1972, 41–42).

17. Pulcini (2003, 89).

18. Neumann (1996).

19. Experience with international politics of course left the Russians with few doubts about the likelihood that other powers would acquiesce to their acquisition of Constantinople, and both the tsar and Nesselrode thus repeatedly disavowed interest in direct acquisition. At the same time, however, Russian plans for the event of Turkish collapse consistently involved Russia temporarily taking control of the Bosporus; as the British experience in Egypt a decade and a half later was to show, of course, "purely temporary" control could become quite protracted.

20. Rich (1985, 14).

21. Curtiss (1979, 62).

22. For examples of studies that present the Japanese decision for war as ultimately self-defeating and irrational, see Snyder (1991, ch. 4) and Kupchan (1994, ch. 5).

23. Ike (1967, e.g., 247–249).

24. The decision that the war could end only through unconditional surrender came quite early in the war, although it was made public only in January 1943. For more information, see chapter 4.

25. Sagan (1988) provides the best extant account of the preventive motivations behind the Japanese attack, including substantially more detail than can be presented here.

26. Quoted in Yagame (2006, 97).

27. Quoted in Sagan (1988, 337). See also Marder (1981, vol. 1, 175).

28. Miwa (1975, 116).

29. Sagan (1988, 326), Snyder (1991, 125–126).

30. For the centrality of resource problems in Japanese strategic planning, see Barnhart (1987).

31. Ike (1967, 238), Marder (1981, vol. 1, 166–167).

32. Marder (1981, vol. 1, 176), Sagan (1988, 342).

33. Ike (1967, 196–200).

34. Ibid., 152.

35. Ibid., 247–249.

36. For a discussion of unconditional surrender that focuses in particular on Japan, see Hellegers (2002).

37. Quoted in Armstrong (1961, 16–17, 19).

38. Sagan (1988, 323).

39. Hellegers (2002, 10).

40. Informational accounts have a particularly difficult time explaining this conflict: by 1944 at the latest it was apparent to all that Japan was doomed, and yet fighting continued, with quite possibly over half of all deaths in the war occurring after this point. Dower (1993, 293–294).

41. Hasegawa (2005).

42. Hellegers (2002, 149–155).

43. Even the acceptance of the figurehead role for the emperor was controversial with large segments of the US public. See for example Kecskemeti (1958, 163–167).

44. Gieling (1999, ch. 6) usefully reviews relevant sermons by high-ranking figures that discussed the decision for peace, including calculated rhetorical shifts in the way that the war was presented to the public.

45. On the context of the 1975 agreement, see Hiro (1989, 15–16).

46. *New York Times*, "Khomeini Dismisses Truce Offer, Vowing a Fight to the End," October 1, 1980, A1.

47. See for example King (1987, 10), Karsh (1987/88, ch. 1), Chubin and Tripp (1988, 28–29), Hiro (1989, 37), and Ghareeb (1990, 35). For an exception that explains the Iraqi invasion (and all other political decisions by either side) in terms of domestic politics, see Pelletiere (1992). The strongest domestic political argument is that Saddam fought the war to protect his own position, rather than to protect Iraq as a whole. While his government was unpopular with many in Iraq, it is also true that a wide constituency, most obviously among the Sunnis but also in other ethnic groups, would have been opposed to his replacement by a theocratic government along the lines of the one in Iran.

48. For examples, see Pelletiere (1992, 32).

49. Hiro (1989, 26–27).

50. Robins (1989 46–47).

51. Hiro (1989, 36).

52. Chubin and Tripp (1988, 33).

53. Hiro (1989, 36–37).

54. Seifzadeh (1997, 94).

55. Robins (1989, 47), Hiro (1989, 38). It is worth noting that domestic funding for the war in Iran and Iraq came primarily from oil revenues, although the Iraqis also benefited from large foreign loans.

56. Hiro (1989, 46).

57. For examples, see Chubin (1989) and Pelletiere (1992).

58. Chubin and Tripp (1988, 40).

59. Iraq of course encouraged rebellion among the minority populations, with the result, for example, that Iranian Kurds were cooperating with Iraq at the same time that Iraqi Kurds were assisting Iran. Among the core population that supported the revolution, however, the regime did not face substantial resistance to its war aims.

60. For examples, see Chubin and Tripp (1988, 38, 49) and Gieling (1999, ch. 6).

61. Chubin and Tripp (1988, 52).

62. Hiro (1989, 243).

63. Quoted in Hiro (1989, 32).

64. Pelletiere (1992, 29) describes the desire to export the revolution violently as the preference of a "radical fringe" that did not control policy after the revolution. Khadduri (1988, 67) notes that Khomeini espoused a peaceful interpretation of jihad prior to the revolution, although he believes that Khomeini switched to a more violent interpretation once in power.

65. This case is admittedly unusual, in that it is almost certainly true that the revolutionary leadership was particularly predisposed to explain behavior in dispositional terms, given its theologically based tendency to divide the world into regions of believers and unbelievers, where the latter were inherently corrupt. This sort of predisposition could easily make the dispositional inference more likely.

66. Chubin (1989).

67. *New York Times*, "An Old Letter Casts Doubts on Iran's Goal for Uranium," October 5, 2006, A14.

6. Short Wars of Optimism

1. On the disputes over loan forgiveness and Kuwaiti overproduction, see Baram (1993) and Khadduri and Ghareeb (1997, chs. 6–7).

2. For a history of this claim, including a significant crisis in 1961 that was resolved only through British intervention, see Finnie (1992).

3. Thus, for example, even with 100,000 Iraqi soldiers massed on the border, the Kuwaiti representative at the final meeting between the two sides—at Jidda two days before the invasion—was instructed to hold firm in the face of pressure. Khadduri and Ghareeb (1997, 115–117).

4. Kostiner (1993, 112–114), Yetiv (2004, 22).

5. Some have argued that Saddam may also have gained confidence from the expectation, grounded in Cold War history, that American opposition would ensure that Iraq would receive Soviet support. Gorbachev (1996, 552), Hassan (1999, 4, 39). Although plausible, there is little evidence that expectations of Soviet support played

a significant role in Iraqi thinking; rapid and categorical Soviet denunciations of Iraqi policy in any event provided no basis for optimism as the crisis unfolded.

6. For general discussions of casualty aversion and its implications for the foreign policy of the United States or of democracies more generally, see Smith (2005) and Wrede-Braden (2007).

7. *New York Times*, "Excerpts from Iraqi Document on Meeting with U.S. Envoy," September 23, 1990, 19. The Iraqi army had demonstrated the ability to inflict significant casualties in the war with Iran, at times killing well over ten thousand Iranians in a single battle. Clodfelter (2007, 627–629).

8. Matar (1981, 95).

9. Quoted in Long (2004, 14). See also comments to this effect by David Newton, who as US ambassador to Iraq in the mid-1980s had Saddam emphasize this point to him on several occasions. Woodward (1991, 258). Analysis of polling data after the war indicates that the expectation that significant casualties would lead to a substantial drop in popular support for the war was "basically sound," although the Iraqis failed to impose the level of casualties necessary to bring about this drop in support. Mueller (1994, 121, 124–129).

10. Simpson (1991, pp 273, 281–282). American policymakers cited this concern as a reason for ending the ground war when they did, although at least one subsequent study found that public opinion demonstrated little concern for Iraqi civilian or military casualties. Bush and Scowcroft (1998, 485), Mueller (1994, 122–123).

11. On constructive engagement, see Rubin (1993) and Karabell (1995).

12. That said, when she observed that "we have no opinion on the Arab-Arab conflicts, like your border disagreement with Kuwait," she relied on a stance that was designed to deal with a situation in which Iraq requested American support in the border dispute. Viorst (1991, 66). The comment was thus more encouraging to Iraq than it had been meant to be, but it was not dramatically out of line with US policy, and it was a reasonable response given that Glaspie had not been given time before the meeting to check in with Washington. Indeed, in contemporary domestic discussions, State Department representatives observed that the United States had "no special defense or security commitments to Kuwait," and Glaspie was never instructed to make unambiguous deterrent threats. Quoted in Stein (1992, 152).

13. As Jervis (1993, 177) notes, simultaneously deterring and reassuring a potential opponent in international politics is an extremely difficult task, and conceivably an impossible one in this case.

14. Schwarzkopf (1992, 385), Knights (2005, 20–25).

15. Bin et al. (1998, 75–78).

16. Viorst (1991, 67).

17. For the desire for an "Arab" solution to the crisis, see for example Khadduri and Ghareeb (1997, 161–167). The expectation that the Saudis might buy off Saddam clearly was not groundless, given their history of using money to resolve conflict; indeed, the Americans worried that the Saudis might decide to resort to financial diplomacy and undermine the strong line that the Bush administration preferred. Yetiv (2004, 34).

18. Woodward (1991, 268), Long (2004, 39–43). The Iraqi claim not to have intended to invade Saudi Arabia gains credence from the observation that they did not attack during the period immediately after the invasion or during the first few weeks of the

American deployment to Saudi Arabia, during which time the force protecting the main Saudi oilfields would have posed little greater obstacle to an Iraqi advance than the Kuwaiti army had.

19. See for example Simpson (1991, 276). Saddam also sought to tie opposition to the Iraqi invasion to Israel, in the hope that doing so would fracture the coalition.

20. *New York Times*, "War and Peace: A Sampling from the Debate on Capital Hill," January 11, 1991, A8. See also *New York Times*, "Bush May Recall Congress to Consider Force in Gulf," November 29, 1990, A1, and *New York Times*, "Legislators Take Sides for Combat or for Reliance on the Sanctions," January 11, 1991, A1.

21. *New York Times*, "Debate This Week: Approval Appears Likely After Heavy Arguing, Say Hill Leaders," January 9, 1991, A6.

22. Bush and Scowcroft (1998, 445–446).

23. For examples of attempts to find ways to credibly signal resolve, see Bush and Scowcroft (1998, 421–423, 426).

24. Viorst (1991, 68).

25. Simpson (1991, 249–251), Woodward (1991, 335–336).

26. Quoted in Yetiv (2004, 165).

27. *New York Times*, "Amid Preparations, No One Can Say When a War Might Break Out or End," October 21, 1990, 12; and Bush and Scowcroft (1998, 425). For examples of higher public estimates at the time, see *New York Times*, "Fighting the Iraqis: Four Scenarios, All Disputed," November 19, 1990, A1, and *New York Times*, "War and Peace: A Sampling from the Debate on Capital Hill," January 11, 1991, A8. A contemporary statistical estimate predicted that total deaths on both sides from the war would exceed 100,000 and might reach 1,000,000. Cioffi-Revilla (1991).

28. Bush and Scowcroft (1998, 428).

29. Khadduri and Ghareeb (1997, 132).

30. Given the intractability of the Israel-Palestine conflict and of Syrian involvement in Lebanon, these demands were seen in the region and elsewhere as a rhetorical ploy to cover an intent to remain in Kuwait indefinitely. Yetiv (2004, 40).

31. Gorbachev (1996, 551–565) describes Soviet peace efforts in detail.

32. For the text of this proposal, see *New York Times*, "Moscow's Statement: Transcript of Comments on Soviet Peace Proposal," February 23, 1991, 5.

33. For details on the cease-fire terms, see Schwarzkopf (1992, 479–490). For Iraqi acceptance of the, in their words, "unfair and vindictive measures" contained in the relevant Security Council Resolutions (in particular Resolution 687), see *New York Times*, "Excerpts From Letter to U.N.: Iraqis 'Accept This Resolution,'" April 8, 1991, A6.

34. Bush and Scowcroft (1998, 489), Yetiv (2004, 219).

35. Quinlivan (1999). Oakes (2006) argues that a leader in Saddam's position is more likely to use repression to retain power than diversion.

36. I am aware of four book-length studies of this war: English (1971), the best English-language source on the conflict; Hunt and Townsend (1858) and Outram (1860), which are most useful for military questions; and Bushev (1959), which provides an alternative (Russian) perspective, albeit based on similar sources as those used by English. Akhmedzhanov (1971) provides a useful history of the Herat question, albeit suffused with a significant anti-British bias. A number of works also discuss various

aspects of the conflict as one part of a broader analysis. Most work unsurprisingly focuses on the British perspective, with little information on decision making inside the Iranian court; in this regard, Amanat (1997, esp. ch. 7) is particularly useful for its insight into the Iranian court. See also Walpole (1912, vol. VI, 266–273).

37. It took more than a month for communications to pass between European capitals and Asian cities like Tehran and Bombay in which significant decisions concerning the war were made. On the difficulty of communication, see for example Hunt and Townsend (1858, esp. ch. 4).

38. For the significance of Iran and Herat within the Great Game, see Akhmedzhanov (1971, 17–21).

39. Outram (1860, 237–238), Barker (1915, 32).

40. A total of roughly 1,500 to 2,000 people died in the war (the vast majority of them Iranian soldiers), a very low total for a war between two relatively large countries.

41. One anecdote provides evidence of the extent to which the Iranians recognized the inevitability of military defeat in the event of war with Britain. At the start of the war, the Iranian prime minister ensured that one of his political enemies would be named as the commander-in-chief for the Iranian forces, thus positioning him to take the blame for the anticipated subsequent military disaster. That man, similarly recognizing the futility of fighting, took so long to reach the front that the war was already lost by the time that he arrived. English (1971, 119–120).

42. Bushev (1959, 40).

43. Quoted in English (1971, 28–29).

44. Waller (1990).

45. See Barker (1915, 2) and English (1971, 63). Instead of invading, the British formed an alliance with the prince of Kabul Dost Mohammed Khan (the man who ultimately reunified Afghanistan in 1863), although the alliance ended up being more of a defensive nature and was in any event not sealed until late January, leaving little time for any real action in the war. Kaye (1864, vol. 1, 427–448).

46. Hunt and Townsend (1858, 180), Walpole (1912, 270–271). Amanat (1997, 278, 286) reports Iranian disappointment when the news of the end of the Crimean War arrived, at a point when the Iranians had already committed to the seizure of Herat.

47. It is worth noting that the possibility for Russian intervention—an additional potential source of divergent expectations—seems to have played no role in the move to war, despite fears in some British quarters that the war with Iran might herald a new clash with Russia. Bushev (1959, 86). The Russians were not eager for another fight with Britain so recently after the Crimean War, and there is no indication that the leaderships on either side ever considered such an eventuality at all likely.

48. The evidence in this paragraph is taken from Amanat (1997, 293–302).

49. English (1971, 138).

50. For the full text of the treaty, see Rawlinson (1875, vol. 4, 370–373).

51. Amanat (1997, 304).

52. See for example English (1971, 43).

53. Bakhash (1978), Martin (2005).

54. Ashley (1879, vol. 2, 346), Chamberlain (1987, 97–99).

55. English (1971, 139), quoted in Bushev (1959, 130).

56. Ibid., 140–141.

57. Holt (1964, 202–204). It is possible that a similar outcome would have arisen over the war with Iran, but in practice the Chinese dispute, in which Britain was attempting to force the sale of an addictive narcotic to address its trade deficit, provided a far more attractive basis for opposition criticism. Hurd (1967).

7. The Limits on Leaders

1. Oakes (2006, 432). For a similar interpretation, see Levy and Vakili (1992).

2. *Time Magazine*, February 14, 1983, 65. On the history of the islands, see Hoffmann and Hoffmann (1984) and Freedman (2005, vol. 1, chs. 1–3). For a summary of what limited strategic significance the islands have, see Sloan (2005).

3. On the Falklands lobby, which had managed as early as 1967 to get a public pledge that no deal would be struck against the islanders' wishes, see Dillon (1989, ch. 3) and Charlton (1989, ch. 4).

4. Freedman (2005, vol. 1, 153–154).

5. For evidence of the strong positive reaction in Argentina, see Femenia (1996, esp. 96–98).

6. For the performance of the junta in the years prior to 1982, see Rock (1985, 366–367), Pion-Berlin (1985), and Vacs (1987).

7. Pion-Berlin (1985), Arquilla and Rasmussen (2001).

8. Vacs (1987, 24–29).

9. Levy and Vakili (1992, 130), Lewis (2001, 146–147).

10. Gamba (1987, 111–112), Freedman (2005, vol. 1, 153).

11. Freedman and Gamba-Stonehouse (1991, 105), Gamba (1987, 115), Freedman (2005, vol. 1, ch. 15).

12. Lebow (1985, 108), Freedman (2005, vol. 1, 159). It was clear that this behavior was intended to strengthen support for the government. See *Latin American Weekly Report*, "Argentina: Islands Used as Vote-Catchers," March 12, 1982, available online at http://www.latinnews.com/arcarticle.asp?articleid=78765.

13. See for example Freedman (2005, vol. 1, 71, 107–109).

14. See for example Hastings and Jenkins (1983, 59–60).

15. For succinct rebuttals to the argument that the junta acted in response to economic unrest, see Gamba (1987, 132) and Arquilla and Rasmussen (2001, 748).

16. For works that emphasize the role of divergent expectations, see for example Lebow (1985), Freedman and Gamba-Stonehouse (1991), and Fravel (2010).

17. Toase (2005, 147–151).

18. Gamba (1987, 137–140), Freedman (2005, vol. 1, 190–191).

19. Thatcher (1993, 177).

20. Freedman (2005, vol. 1, 57). On trip-wire forces more generally, see Schelling (1966, 47). Even after the invasion, British decision makers continued to insist that a larger standing commitment prior to the war simply was not an option. Nott (2005, 58).

21. The British also planned to withdraw the survey team that provided the only British presence on South Georgia, although as with the *Endurance* this decision was forestalled at the last minute.

22. See for example Freedman (2005, vol. 1, e.g. 116–118).

23. Ibid., vol. 1, ch. 11.

24. Franks (1992, 26–27, 32).

25. Gustafson (1988, 62–63), Freedman and Gamba-Stonehouse (1991, 78), Freedman (2005, vol. 1, 153).

26. Freedman and Gamba-Stonehouse (1991, 82–83, 107).

27. Gamba (1987, 145–146).

28. *The Times of London*, "Galtieri: No Regrets, No Going Back," June 12, 1982, 4.

29. Quoted in Freedman and Gamba-Stonehouse (1991, 200).

30. Arquilla and Rasmussen (2001, 754–758).

31. Freedman and Gamba-Stonehouse (1991, 257, 273, 277).

32. See for example Hughes and Larson (1985) and Arquilla and Rasmussen (2001).

33. Thatcher (1993, 173–174).

34. Freedman and Gamba-Stonehouse (1991, 241).

35. Hastings and Jenkins (1983, 123–124), Freedman (2005, vol. 2, 69–82).

36. Freedman (2005, vol. 1, 206, 211).

37. Nott (2005, 59).

38. Freedman (2005, vol. 1, 143–147, 168–169, 199–200). There were also concerns that there might be a British submarine in the area.

39. Ibid., vol. 1, 187.

40. For the full history, see Haig (1984, ch. 13) and Freedman (2005, vol. 2, sections 2 and 4). The talks included a lot of detailed bargaining over subsidiary issues, in large part because the mediators sought to build a consensus by starting with issues—such as the composition of an interim authority—on which agreement might be reached before tackling the more difficult questions of sovereignty, while each side was reluctant to directly reject any mediator proposals to avoid being blamed for the failure of negotiations. On the central sovereignty question, however, neither side ever made a substantial concession.

41. The British were permitted to make use of an American base on (British-owned) Ascension Island, and the United States and other countries (notably France) provided intelligence on Argentine capabilities. That said, the Argentine Navy ultimately effectively withdrew from the war in part because naval leaders believed that the United States was providing Britain with extensive satellite intelligence, well beyond the very limited amount that was actually provided. Freedman and Gamba-Stonehouse (1991, 270), Freedman (2005, vol. 2, 71, 236).

42. Freedman (2005, vol. 2, e.g. 173–174).

43. Lebow (1985, 109), *Latin American Weekly Report*, "The Forces that Galtieri Unleashed," April 30, 1982, available online at http://www.latinnews.com/arcarticle. asp?articleid=79445. See also Freedman and Gamba-Stonehouse (1991, 313).

44. Quoted in Freedman (2005, vol. 2, 544).

45. *The Times of London*, "Galtieri: No Regrets, No Going Back," June 12, 1982, 4.

46. Freedman (2005, vol. 2, 511–512).

47. Attacks on the Argentine mainland would have been politically counterproductive. Thatcher (1993, 221).

48. Vacs (1987, 29).

49. *The Times of London*, "Military Defeat Hammers Last Nail into Galtieri's Political Coffin," June 18, 1982, available online at http://www.latinnews.com/arcarticle. asp?articleid=80018.

50. Arquilla and Rasmussen (2001).

51. Gamba (1987, 181).

52. For a similar argument about the limits of diversionary war, see Fravel (2010).

53. Historiography on this conflict is quite limited, and accounts of specific incidents are frequently contradictory. Indeed, Zeidner (2005, 3), in an admittedly opinionated work on the war, asserts that "much, if not most, of what [prior scholars] have said about it is wrong." The best-covered aspect is the great power diplomacy of the period, in which the relationship between France and Britain is particularly important. There is also a reasonable literature on the emergence of the Kemalist movement to replace the Ottoman Porte and on the partially coterminous Greco-Turkish War. Coverage of events on the ground in Cilicia and of direct negotiation between the two sides is particularly limited. Of the four sources that treat the war in detail, Saakian (1986) and Tachjian (2004) are most interested in the fate of the Armenian population and are consequently somewhat less useful for the questions here. Nakache (1999) is the most detailed source, but her work relies almost exclusively on French-language sources. Zeidner (2005) is thus the only significant work to look in detail at the political and military developments of the war using both French and Turkish sources. There are, of course, also works in Turkish, which my language skills do not permit me to use; even here, however, there apparently is little in the way of serious scholarly work. Zeidner (2005, 3–4).

54. Technically, the French possessions in territory claimed by Turkey included some land east of the Amanus Mountains that typically is not considered part of Cilicia. While the greater difficulty of projecting power further from the coast meant that there were significant differences between the two regions with respect to the military course of the war, in political negotiations the distinction was rarely significant. Nakache (1999, part II). References to Cilicia here thus should be read as including the eastern territory.

55. The map was produced by the Palestinian Academic Society for the Study of International Affairs and is available on a free content license at http://en.wikipe dia.org/wiki/Image:Sykes-Picot-1916.gif.

56. Toynbee (1922, 88–89). Thus, for example, the French estimated that they held some 75 percent of the entire Ottoman public debt, while independent estimates put the figure at more than 60 percent; they were thus hopeful that the empire might survive, if in a diminished form. Busch (1976, 200), Saakian (1986, 41). For a detailed discussion of French investments in Syria and Cilicia in particular, see Nakache (1999, 51–55, 75–84).

57. Andrew and Kanya-Forstner (1974, 82), Cumming (1938). For useful discussions of Allied negotiations over the postwar disposition of the Ottoman Empire, see Howard (1931), Andrew and Kanya-Forstner (1974), and Tanenbaum (1978).

58. Zeidner (2005, 153–156).

59. Sonyel (1975).

60. Zeidner (2005, 78–80, 139, 177–178).

61. A truce reached on May 30, 1920, was never fully implemented and ultimately collapsed after only a few weeks. Zeidner (2005, 238).

62. Zeidner (2005, 186).

63. Tachjian (2004, 168).

64. The importance of the colonial party is a consistent theme in literature on French policy in Cilicia. In addition to the works cited below by Andrew and Kanya-Forstner, who directly address the colonial party, see for example Saakian (1986, 48–49), Nakache (1999, 89–95), Tachjian (2004, 21–26), and Zeidner (2005, 22–31).

65. Andrew (1976, 145).

66. Andrew and Kanya-Forstner (1981), Abrams and Miller (1976, 686).

67. Andrew and Kanya-Forstner (1974, 96), Zeidner (2005, 22–29).

68. Andrew and Kanya Forstner (1974, 85).

69. Andrew and Kanya Forstner (1976, 991–993), Zeidner (2005, 258). Andrew and Kanya-Forstner do note that members of the colonial party were not above using expeditions to further both French prestige and their own business interests simultaneously. The business side was typically secondary, however, and in any event it is hard to argue that France as a whole benefited from policies designed to serve the specific economic interests of policymakers.

70. Andrew and Kanya-Forstner (1976, 986). The one exception was with French silk producers, as French sericulture had been effectively destroyed by disease. Silk production played only a minor role in the French economy, however.

71. As a side note, this disjuncture poses a significant problem for Marxist interpretations of French imperialism such as that of Saakian (1986). For a relevant discussion, see the debate between Abrams and Miller (1976) and Andrew and Kanya-Forstner (1976), in which the latter are far more convincing. See also Andrew and Kanya-Forstner (1981, 17).

72. See Nakache (1999, 725) for a parliamentary speech by Édouard Daladier, a future prime minister, that frames the war in Cilicia in precisely these terms.

73. The French do seem to have held out some hope for a security guarantee from the United States, but the return to isolationism in Washington, especially following President Wilson's stroke in October 1919, killed that prospect. To the extent that a guarantee ever was a possibility, however, French colonial designs, conflicting as they did with the spirit of Wilson's Fourteen Points, were again a significant irritant. Zeidner (2005, 55).

74. Cumming (1938, 13–20), Tanenbaum (1978, 6–8).

75. Cumming (1938), McCrum (1978).

76. Quoted in Andrew and Kanya-Forstner (1976, 984), my translation.

77. Zeidner (2005, 139), Nakache (1999, 627–640).

78. Nakache (1999, 383).

79. Ibid., 717, 721.

80. Zeidner (2005, 262). See Nakache (1999, 707–709) for a discussion of coverage of Cilicia in the press, including a list of specific articles, many of which had first been published in British newspapers.

81. Complaints increased over the course of 1920, and by the end of the year parliament was refusing to renew funding for more than an additional two months. Nakache (1999, 726).

82. Zeidner (2005, 273).

83. Tachjian (2004, 109, 136).

84. Khoury (1987, esp. part III).

85. Zeidner (2005, 255).

86. Aharonian (1964, 68–69), Zeidner (2005, 256).

87. Zeidner (2005, 272).

88. Khoury (1987, 46).

89. Some readers may wonder whether Turkish domestic politics similarly influenced the duration and severity of the war, given the strong split between the Porte and the Kemalists. In practice, however, the Porte's decision not to resist was out of line with the preferences of the public that it claimed to represent, but in a direction that favored peace rather than war; by contrast, the Kemalist stance seems to have matched the preferences of the Turkish public fairly closely. Zürcher (1984, 116).

90. Sonyel (1975, 23–24).

91. Indeed, the only direct reference that I found to such concerns is Toynbee (1922, 85), who complained specifically about lack of French concern about the possibility that Turkey might seek to undermine the French position in Syria.

92. The significance of Ottoman debts can be seen in the intensity of Anglo-French debates over whether reparations should have priority over prewar debts. Montgomery (1972, 779).

93. Zeidner (2005, 267).

94. Nakache (1999, 738), Zeidner (2005, 256).

95. Burrows (1986), Zeidner (2005, 123–126), Saakian (1986, 44–45).

96. Zeidner (2005, 125–126).

97. Saakian (1986, 41), Tachjian (2004, 115–116). The early French recognition contrasted especially strongly with the British failure to recognize the growing irrelevance of the old Ottoman government, whom they repeatedly blamed for violence carried out by Kemalist forces over whom that government had no control. Zeidner (2005, 185).

98. Aharonian (1962, 5–6), Zeidner (2005, 238, 246).

99. Andrew and Kanya-Forstner (1981, 5).

Conclusion

1. On spoiler problems, see Stedman (1997).

2. See especially Fearon (2004).

3. Walter (1997, 2002).

4. See for example Collier et al. (2004) and Ross (2004).

5. For a representative example that also highlights the reluctance of the international community, and the United States in particular, to intervene even once conflict begins, see Power (2002).

6. On the logical impossibility of perfectly predicting war *ex ante*, see Gartzke (1999).

7. Kydd (2003, 2006).

8. Mearsheimer (1990), Fukuyama (1992).

9. Mueller (1989, 2004).

10. This discussion assumes, of course, that China's development continues along its current course. Japan's experience of tremendous growth in the 1980s followed by a decade of stagnation provides a salutary reminder that current trends are not guaranteed to continue, and concerns about an overheated economy and about internal unrest associated with increasing economic inequality and unrepresentative government provide potential bases for concern. That said, China's large population and other advantages provide reason to believe that, even should its growth be derailed for any of these reasons, the country will ultimately rise to rival the United States as a world power.

11. See for example Friedberg (2011).

12. It is worth noting, however, that in recent history, internal unrest in China has been associated with *increased* flexibility in external disputes. Fravel (2005).

13. On this point, see the discussion of the stability-instability paradox in Snyder and Diesing (1977).

Bibliography

Abente, Diego. 1987. "The War of the Triple Alliance: Three Explanatory Models." *Latin American History Review* 22(2): 47–69.

Abrams, L., and D. J. Miller. 1976. "Who Were the French Colonialists? A Reassessment of the Parti Colonial, 1890–1914." *Historical Journal* 19(3): 685–725.

Aharonian, Avetis. 1962. "From Sardarapat to Sevres and Lausanne (Part 1)." *Armenian Review* 15 (3–59): 3–13.

——. 1964. "From Sardarapat to Sevres and Lausanne (Part 6)." *Armenian Review* 17 (1–65): 64–73.

Akhmedzhanov, G. A. 1971. *Geratskii Vopros v XIX Veke*. Tashkent: Fan.

Allen, Martin. 2005. *Himmler's Secret War: The Covert Peace Negotiations of Heinrich Himmler*. London: Robson Books.

Amanat, Abbas. 1997. *Pivot of the Universe: Nasr al-Din Shah Qajar and the Iranian Monarchy, 1831–1896*. Berkeley: University of California Press.

Anderson, Thomas P. 1981. *The War of the Dispossessed: Honduras and El Salvador, 1969*. Lincoln: University of Nebraska Press.

Andrew, C. M. 1976. "The French Colonialist Movement during the Third Republic: The Unofficial Mind of Imperialism." *Transactions of the Royal Historical Society* 5th Ser., 26: 143–166.

Andrew, C. M., and A. S. Kanya-Forstner. 1974. "The French Colonial Party and French Colonial War Aims, 1914–1918." *Historical Journal* 17(1): 79–106.

——. 1976. "French Business and the French Colonialists." *Historical Journal* 19(4): 981–1000.

——. 1981. *The Climax of French Imperial Expansion, 1914–1924*. Stanford: Stanford University Press.

Annual Register, or a View of the History and Politics of the Year 1854. 1855. London: F & J Rivington.

Argyll, George Douglas, Eighth Duke of. 1906. *Autobiography and Memoirs*. London: John Murray.

Armstrong, Anne. 1961. *Unconditional Surrender: The Impact of the Casablanca Policy upon World War II*. New Brunswick, NJ: Rutgers University Press.

Arquilla, John, and María Moyano Rasmussen. 2001. "The Origins of the South Atlantic War." *Journal of Latin American Studies* 33(4): 739–775.

Ashley, Evelyn. 1879. *The Life and Correspondence of Henry John Temple, Viscount Palmerston*. London: R. Bentley.

Aumann, Robert J. 1976. "Agreeing to Disagree." *Annals of Statistics* 4(6): 1236–1239.

[261]

Ayache, Germain. 1981. *Les Origines de la Guerre du Rif.* Paris: Publications de la Sorbonne.

Baker, William D., and John R. Oneal. 2001. "Patriotism or Opinion Leadership? The Nature and Origins of the "Rally 'Round the Flag" Effect." *Journal of Conflict Resolution* 45(5): 661–687.

Bakhash, Shaul. 1978. *Iran: Monarchy, Bureaucracy, and Reform under the Qajars, 1858–1896.* London: Ithaca Press for the Middle East Centre, St. Antonys College.

Baldwin, David A. 1989. *Paradoxes of Power.* New York: Basil Blackwell.

Baldwin, Hanson W. 1950. *Great Mistakes of the War.* New York: Harper & Brothers.

Balfour, Michael. 1970. "Another Look at Unconditional Surrender." *International Affairs* 46(4): 719–736.

Balfour, Sebastian. 2002. *Deadly Embrace: Morocco and the Road to the Spanish Civil War.* Oxford: Oxford University Press.

Baram, Amatzia. 1993. "The Iraqi Invasion of Kuwait: Decision-making in Baghdad." In *Iraq's Road to War*, edited by Amatzia Baram and Barry Rubin, 5–36. New York: St. Martin's Press.

Barker, George Digby. 1915. *Letters from Persia and India, 1857–1859: A Subaltern's Experiences in War.* London: G. Bell and Sons.

Barman, Roderick J. 1999. *Citizen Emperor: Pedro II and the Making of Brazil, 1825–91.* Stanford: Stanford University Press.

Barnes, James J., and Patience P. Barnes. 1980. *Hitler's 'Mein Kampf' in Britain and America: A Publishing History, 1930–1939.* Cambridge: Cambridge University Press.

Barnhart, Michael A. 1987. *Japan Prepares for Total War: The Search for Economic Security, 1919–1941.* Ithaca: Cornell University Press.

Bator, Francis M. 2008. "No Good Choices: LBJ and the Vietnam/Great Society Connection." *Diplomatic History* 32(3): 309–340.

Bell, P.M.H. 1974. *A Certain Eventuality: Britain and the Fall of France.* London: Saxon House.

——. 1990. *John Bull and the Bear: British Public Opinion, Foreign Policy and the Soviet Union, 1941–1945.* London: Edward Arnold.

Bennett, D. Scott, and Allan C. Stam III. 1996. "The Duration of Interstate Wars, 1816–1985." *American Political Science Review* 90(2): 239–257.

——. 1998. "The Declining Advantages of Democracy: A Combined Model of War Outcomes and Duration." *Journal of Conflict Resolution* 42(3): 344–366.

Bensahel, Nora, Olga Oliker, Keith Crane, Richard R. Brennan, Jr., Heather S. Gregg, Thomas Sullivan, and Andrew Rathmell. 2008. *After Saddam: Prewar Planning and the Occupation of Iraq.* Santa Monica, CA: RAND Corporation.

Bergan, Daniel E. 2009. "The Draft Lottery and Attitudes towards the Vietnam War." *Political Opinion Quarterly* 73(2): 379–384.

Berghahn, Volker. 1976. "Naval Armaments and Social Crisis: Germany before 1914." In *War, Economy, and the Military Mind*, edited by Geoffrey Best and Andrew Wheatcroft, 61–88. London: Croom Helm.

Berinsky, Adam J. 2009. *In Time of War: Understanding American Public Opinion from World War II to Iraq.* Chicago: University of Chicago Press.

Berman, Larry. 1982. *Planning a Tragedy: The Americanization of the War in Vietnam.* New York: W. W. Norton.

Bernstein, Harry. 1973. *Dom Pedro II.* New York: Twayne.

Bethell, Leslie. 1996. *The Paraguayan War (1864–1870).* London: Institute of Latin American Studies.

[262]

Bialer, Uri. 1980. *The Shadow of the Bomber: The Fear of Air Attack and British Politics, 1932–1939*. London: Royal Historical Society.

Biddle, Stephen D. 2004. *Military Power: Explaining Victory and Defeat in Modern Battle*. Princeton: Princeton University Press.

Bin, Alberto, Richard Hill, and Archer Jones. 1998. *Desert Storm: A Forgotten War*. Westport, CT: Praeger.

Blainey, Geoffrey. 1988. *The Causes of War*. New York: Free Press.

Blumberg, Arnold. 1990. *A Carefully Planned Accident: The Italian War of 1859*. Selinsgrove, PA: Susquehanna University Press.

Box, Pelham Horton. 1929. *The Origins of the Paraguayan War*. Urbana: University of Illinois Press.

Box-Steffensmeier, Janet, and Bradford S. Jones. 2004. *Event History Modeling: A Guide for Social Scientists*. New York: Cambridge University Press.

Broszat, Martin. 1966. *German National Socialism, 1919–1945*. Santa Barbara: CLIO Press.

———. 1970. "Soziale Motivation und Führer-Bindung des Nationalsozialismus." *Vierteljahrshefte für Zeitgeschichte* 18(4): 392–409.

Browning, Christopher. 1992. *Ordinary Men: Reserve Police Battalion 101 and the Final Solution in Poland*. New York: HarperCollins.

Bucholz, Arden. 2001. *Moltke and the German Wars, 1864–1871*. New York: Palgrave.

Bueno de Mesquita, Bruce. 1981. *The War Trap*. New Haven: Yale University Press.

Bueno de Mesquita, Bruce, Alastair Smith, Randolph M. Siverson, and James D. Morrow. 2003. *The Logic of Political Survival*. Cambridge: MIT Press.

Bulloch, John, and Harvey Morris. 1989. *The Gulf War: Its Origins, History and Consequences*. London: Methuen.

Bullock, Alan. 1953. *Hitler: A Study in Tyranny*. New York: Harper & Row.

Burridge, Trevor. 1985. *Clement Attlee: A Political Biography*. London: Jonathan Cape.

Burrows, Mathew. 1986. " 'Mission Civilisatrice': French Cultural Policy in the Middle East, 1860–1914." *Historical Journal* 29(1): 109–135.

Busch, Briton Cooper. 1976. *Mudros to Lausanne: Britain's Frontier in West Asia, 1918–1923*. Albany: State University of New York Press.

Bush, George, and Brent Scowcroft. 1998. *A World Transformed*. New York: Alfred A. Knopf.

Bushev, P. P. 1959. *Gerat i Anglo-Iranskaia Voina, 1856–1857 gg*. Moscow: Izdatel'stvo Vostochnoi Literatury.

Butler, J.R.M. 1957. *Grand Strategy, Volume II: September 1939–June 1941*. London: Her Majesty's Stationary Office.

Cappella, Rosella. 2012. *The Political Economy of War Finance*. Ph.D. diss., University of Pennsylvania.

Carlton, David. 1993. "Churchill in 1940: Myth and Reality." *World Affairs* 156(2): 97–103.

Carroll, Wallace. 1948. *Persuade or Perish*. Boston: Houghton Mifflin.

Casey, Steven. 2001. *Cautious Crusade: Franklin D. Roosevelt, American Public Opinion, and the War against Nazi Germany*. Oxford: Oxford University Press.

Caven, Brian. 1980. *The Punic Wars*. New York: St. Martin's Press.

Cecil, Robert. 1975. *Hitler's Decision to Invade Russia, 1941*. London: Davis-Poynter.

Cederman, Lars-Erik. 2003. "Modeling the Size of Wars: From Billiard Balls to Sandpiles." *American Political Science Review* 97(1): 135–150.

Chamberlain, Muriel E. 1987. *Lord Palmerston*. Washington, DC: Catholic University of America Press.

Chapman, Terrence L., and Dan Reiter. 2004. "The United Nations Security Council and the Rally 'Round the Flag Effect." *Journal of Conflict Resolution* 48(6): 886–909.

Charlton, Michael. 1989. *The Little Platoon: Diplomacy and the Falklands Dispute*. Oxford: Basil Blackwell.

Charmley, John. 1989. *Chamberlain and the Lost Peace*. London: Hodder & Stoughton.

——. 1993. *Churchill: The End of Glory*. London: Hodder & Stoughton.

Chase, John L. 1955. "Unconditional Surrender Reconsidered." *Political Science Quarterly* 70(2): 258–279.

Chief of Council for the Prosecution of Axis Criminality. 1946. *Nazi Conspiracy and Aggression*. Washington, DC: U.S. Government Printing Office.

Chiozza, Giacomo. 2002. "Is There a Clash of Civilizations? Evidence from Patterns of International Conflict Involvement, 1946–97." *Journal of Peace Research* 39(6): 711–734.

Chiozza, Giacomo, and Hein E. Goemans. 2003. "Peace through Insecurity: Tenure and International Conflict." *Journal of Conflict Resolution* 47(4): 443–467.

Chubin, Shahram. 1989. "Iran and the War: From Stalemate to Ceasefire." In *The Gulf War: Regional and International Dimensions*, edited by Hanns W. Maull and Otto Pick, 5–16. New York: St. Martin's Press.

Chubin, Shahram, and Charles Tripp. 1988. *Iran and Iraq at War*. Boulder, CO: Westview Press.

Churchill, Winston S. 1950. *The Second World War: The Grand Alliance*. Boston: Houghton Mifflin.

Cioffi-Revilla, Claudio. 1991. "On the Likely Magnitude, Extent, and Duration of an Iraq-US War." *Journal of Conflict Resolution* 35(3): 387–411.

Clark, Alan. 1965. *Barbarossa: The Russian-German Conflict, 1941–1945*. New York: W. Morrow.

Clausewitz, Carl von. [1832] 1976. *On War*. Princeton: Princeton University Press.

Clodfelter, Micheal. 2007. *Warfare and Armed Conflicts: A Statistical Reference to Casualty and Other Figures, 1500–2000*. Jefferson, NC: McFarland.

Coe, Andrew J. 2011. "Costly Peace: A New Rationalist Explanation for War." Unpublished article, Harvard University.

Colaresi, Michael. 2007. "The Benefit of the Doubt: Testing an Informational Theory of the Rally Effect." *International Organization* 61(1): 99–144.

Collier, Paul, Anke Hoeffler, and Måns Söderbom. 2004. "On the Duration of Civil War." *Journal of Peace Research* 41(3): 253–273.

Compton, James V. 1967. *The Swastika and the Eagle: Hitler, the United States, and the Origins of World War II*. Boston: Houghton Mifflin.

Coox, Alvin D. 1977. *The Anatomy of a Small War: The Soviet-Japanese Struggle for Chankufeng/Khasan, 1938*. Westport, CT: Greenwood Press.

——. 1985. *Nomonhan: Japan against Russia, 1939*. Stanford: Stanford University Press.

Copeland, Dale C. 2000. *The Origins of Major War*. Ithaca: Cornell University Press.

Coser, Lewis A. 1956. *The Functions of Social Conflict*. Glencoe, IL: Free Press.

Costello, John. 1991. *Ten Days that Saved the West*. London: Bantam Press.

Croco, Sarah E. 2011. "The Decider's Dilemma: Leader Culpability, War Outcomes, and Domestic Punishment." *American Political Science Review* 105(3): 457–477.

Cumming, Henry H. 1938. *Franco-British Rivalry in the Post-War Near East: The Decline of French Influence*. London: Oxford University Press.

Cunningham, David E. 2006. "Veto Players and Civil War Duration." *American Journal of Political Science* 50(4): 875–892.

Cunninghame Graham, R. B. 1933. *Portrait of a Dictator: Francisco Solano Lopez (Paraguay, 1865–1870)*. London: Heinemann.

Curtiss, John Shelton. 1979. *Russia's Crimean War*. Durham, NC: Duke University Press.

Dallek, Robert. 1979. *Franklin D. Roosevelt and American Foreign Policy, 1932–1945.* New York: Oxford University Press.

Darilek, Richard E. 1976. *A Loyal Opposition in Time of War: The Republican Party and the Politics of Foreign Policy from Pearl Harbor to Yalta.* Westport, CT: Greenwood Press.

Davies, Norman. 1972. *White Eagle, Red Star: The Polish-Soviet War, 1919–20.* New York: St. Martin's Press.

Decalo, Samuel. 1989. *Psychoses of Power: African Personal Dictatorships.* Boulder, CO: Westview Press.

de Figueiredo, Rui J. P., and Barry Weingast. 1999. "The Rationality of Fear: Political Opportunism and Ethnic Conflict." In *Civil Wars, Insecurity, and Intervention*, edited by Barbara Walter and Jack Snyder, 261–302. New York: Columbia University Press.

de la Fuente, Ariel. 2004. "Federalism and Opposition to the Paraguayan War in the Argentine Interior: La Rioja, 1865–67." In *I Die with My Country: Perspectives on the Paraguayan War, 1864–1870*, edited by Hendrik Kraay and Thomas Whigham, 140–153. Lincoln: University of Nebraska Press.

DeRouen, Karl R., Jr., and David Sobek. 2004. "The Dynamics of Civil War Duration and Outcome." *Journal of Peace Research* 41(3): 303–320.

de Soysa, Indra, John R. Oneal, and Yong-Hee Park. 1997. "Testing Power-Transition Theory Using Alternative Measures of National Capabilities." *Journal of Conflict Resolution* 41(4): 509–528.

Dilks, David. 1978. "The Twilight War and the Fall of France: Chamberlain and Churchill in 1940." *Transactions of the Royal Historical Society* Fifth Series, 28: 61–86.

Dillon, G. M. 1989. *The Falklands, Politics and War.* New York: St. Martin's Press.

Doenecke, Justus D. 2000. *Storm on the Horizon: The Challenge to American Intervention, 1939–1941.* New York: Rowman & Littlefield.

Domarus, Max. 1997. *Hitler: Speeches and Proclamations, 1932–1945.* London: I. B. Tauris.

Dower, John W. 1993. *War without Mercy: Race and Power in the Pacific War.* New York: Pantheon Books.

Edwards, Robert. 2006. *White Death: Russia's War on Finland, 1939–40.* London: Weidenfeld & Nicolson.

Eisenhardt, Kathleen M. 1989. "Agency Theory: An Assessment and Review." *Academy of Management Review* 14(1): 57–74.

Ellsberg, Daniel. 1971. "The Quagmire Myth and the Stalemate Machine." *Public Policy* 19: 217–273.

English, Barbara. 1971. *John Company's Last War.* London: Collins.

Ensor, R.C.K. 1939. " 'Mein Kampf' and Europe." *International Affairs (Royal Institute of International Affairs 1931–1939)* 18(4): 478–496.

Farnham, Barbara Rearden. 1997. *Roosevelt and the Munich Crisis: A Study of Political Decision-Making.* Princeton: Princeton University Press.

Fazal, Tanisha M. 2007. *State Death: The Politics and Geography of Conquest, Occupation, and Annexation.* Princeton: Princeton University Press.

Fazal, Tanisha M., V. Page Fortna, Jessica Stanton, and Alex Weisiger. 2006. "The War Initiation and Termination (WIT) Data Set." Paper presented at the American Political Science Association Annual Meeting.

Fearon, James D. 1995. "Rationalist Explanations for War." *International Organization* 49(3): 379–414.

———. 1996. "Bargaining over Objects that Influence Future Bargaining Power." Unpublished manuscript, Stanford University.

[265]

———. 2004. "Why Do Some Civil Wars Last So Much Longer Than Others?" *Journal of Peace Research* 41(3): 275–301.

Femenia, Nora A. 1996. *National Identity in Times of Crises: The Scripts of the Falklands-Malvinas War.* Commack, NY: Nova Science.

Fey, Mark, and Kristopher W. Ramsay. 2007. "Mutual Optimism and War." *American Journal of Political Science* 51(4): 738–754.

Filson, Darren, and Suzanne Werner. 2002. "A Bargaining Model of War and Peace: Anticipating the Onset, Duration, and Outcome of War." *American Journal of Political Science* 46(4): 819–838.

———. 2004. "Bargaining and Fighting: The Impact of Regime Type on War Onset, Duration, and Outcomes." *American Journal of Political Science* 48(2): 296–313.

Fine, Jason P., and Robert J. Gray. 1999. "A Proportional Hazards Model for the Subdistribution of a Competing Risk." *Journal of the American Statistical Association* 94(446): 496–509.

Finnie, David H. 1992. *Shifting Lines in the Sand: Kuwait's Elusive Frontier with Iraq.* Cambridge, MA: Harvard University Press.

Foot, M.R.D. 1966. *SOE in France: An Account of the Work of the British Special Operations Executive in France, 1940–1944.* London: Her Majesty's Stationary Office.

Fornos Peñalba, José Alfredo. 1982. "Draft Dodgers, War Resisters, and Turbulent Gauchos: The War of the Triple Alliance against Paraguay." *Americas* 38(4): 463–479.

Foster, H. Schuyler. 1983. *Activism Replaces Isolationism: U.S. Public Attitudes, 1940–1975.* Washington, DC: Foxhall Press.

Franks, Oliver. 1992. *The Franks Report: Falkland Islands Review.* London: Pimlico.

Fravel, M. Taylor. 2005. "Regime Insecurity and International Cooperation: Explaining China's Compromises in Territorial Disputes." *International Security* 30(2): 46–83.

———. 2010. "The Limits of Diversion: Rethinking Internal and External Conflict." *Security Studies* 19(2): 307–341.

Freedman, Lawrence. 2005. *The Official History of the Falklands Campaign.* London: Routledge.

Freedman, Lawrence, and Virginia Gamba-Stonehouse. 1991. *Signals of War: The Falklands Conflict of 1982.* Princeton: Princeton University Press.

Friedberg, Aaron L. 2011. *A Contest for Supremacy: China, America, and the Struggle for Mastery in Asia.* New York: W. W. Norton.

Friedman, George, and Meredith LeBard. 1991. *The Coming War with Japan.* New York: St. Martin's Press.

Frieser, Karl-Heinz. 2005. *The Blitzkrieg Legend: The 1940 Campaign in the West.* Annapolis, MD: Naval Institute Press.

Frost, Robert I. 2000. *The Northern Wars: War, State, and Society in Northeastern Europe, 1558–1721.* Harlow, England: Longman.

Fukuyama, Frances. 1992. *The End of History and the Last Man.* New York: Free Press.

Gamba, Virginia. 1987. *The Falklands/Malvinas War: A Model for North-South Crisis Prevention.* Boston: Allen & Unwin.

Gartner, Scott Sigmund. 1997. *Strategic Assessment in War.* New Haven: Yale University Press.

———. 2008. "Ties to the Dead: Connections to Iraq War and 9/11 Casualties and Disapproval of the President." *American Sociological Review* 73(4): 690–695.

Gartzke, Erik. 1999. "War Is in the Error Term." *International Organization* 53(3): 567–587.

Gaubatz, Kurt Taylor. 1991. "Electoral Cycles and War." *Journal of Conflict Resolution* 35(2): 212–244.

Genova, B.K.L., and Bradley S. Greenberg. 1979. "Interests in News and the Knowledge Gap." *Public Opinion Quarterly* 43(1): 79–91.

Ghareeb, Edmund. 1990. "The Roots of Crisis: Iraq and Iran." In *The Persian Gulf War: Lessons for Strategy, Law, and Diplomacy*, edited by Christopher C. Joyner, 21–38. Westport, CT: Greenwood Press.

Gieling, Saskia. 1999. *Religion and War in Revolutionary Iran*. London: I. B. Tauris.

Gilbert, Martin. 1977. *Winston S. Churchill, Volume. V: The Prophet of Truth, 1922–1939*. Boston: Houghton Mifflin.

———. 1993. *The Churchill War Papers, Volume I: At the Admiralty, September 1939–May 1940*. London: Heinemann.

———. 1994. *The Churchill War Papers, Volume II: Never Surrender, May 1940–December 1940*. London: Heinemann.

———. 2000. *The Churchill War Papers, Volume III: The Ever-Widening War, 1941*. New York: W. W. Norton.

Gilbert, Martin, and Richard Gott. 1963. *The Appeasers*. London: Weidenfeld and Nicolson.

Gilpin, Robert. 1981. *War and Change in World Politics*. New York: Cambridge University Press.

Goda, Norman J. W. 1998. *Tomorrow the World: Hitler, Northwest Africa, and the Path toward America*. College Station: Texas A&M University Press.

Goddard, Stacie. 2006. "Uncommon Ground: Indivisible Territory and the Politics of Legitimacy." *International Organization* 60(1): 35–68.

Goemans, Hein E. 2000. *War and Punishment: The Causes of War Termination and the First World War*. Princeton: Princeton University Press.

Goldfrank, David M. 1994. *The Origins of the Crimean War*. London: Longman.

Goldhagen, Daniel. 1996. *Hitler's Willing Executioners: Ordinary Germans and the Holocaust*. New York: Alfred A. Knopf.

Gorbachev, Mikhail. 1996. *Memoirs*. New York: Doubleday.

Gorodetsky, Gabriel. 1999. *Grand Delusion: Stalin and the German Invasion of Russia*. New Haven: Yale University Press.

Gustafson, Lowell S. 1988. *The Sovereignty Dispute over the Falkland (Malvinas) Islands*. New York: Oxford University Press.

Györkei, Jenõ, and Miklós Horváth. 1999. *Soviet Military Intervention in Hungary, 1956*. Budapest: Central European University Press.

Haas, Mark L. 2005. *The Ideological Origins of Great Power Politics, 1789–1989*. Ithaca: Cornell University Press.

Haig, Alexander M. Jr. 1984. *Caveat: Realism, Reagan, and Foreign Policy*. New York: Macmillan.

Halder, Franz. 1988. *The Halder War Diary, 1939–1942*. Novato, CA: Presidio.

Hasegawa, Tsuyoshi. 2005. *Racing the Enemy: Stalin, Truman, and the Surrender of Japan*. Cambridge: Harvard University Press.

Hassan, Hamdi A. 1999. *The Iraqi Invasion of Kuwait: Religion, Identity and Otherness in the Analysis of War and Conflict*. London: Pluto Press.

Hassner, Ron. 2004. "To Halve and to Hold: Conflicts over Sacred Space and the Problem of Indivisibility." *Security Studies* 12(4): 2–33.

Hastings, Max, and Simon Jenkins. 1983. *The Battle for the Falklands*. New York: W. W. Norton.

Hauner, Milan. 1978. "Did Hitler Want a World Dominion?" *Journal of Contemporary History* 13(1): 15–32.

Hellegers, Dale M. 2002. *We, the Japanese People: World War II and the Origins of the Japanese Constitution.* Stanford: Stanford University Press.

Henderson, Errol A., and Richard Tucker. 2001. "Clear and Present Strangers: The Clash of Civilizations and International Conflict." *International Studies Quarterly* 45(2): 317–338.

Hendrickson, Ryan C. 2002. "Clinton's Military Strikes in 1998: Diversionary Uses of Force?" *Armed Forces and Society* 28(2): 309–332.

Hildebrand, Klaus. 1970. *The Foreign Policy of the Third Reich.* Berkeley: University of California Press.

Hillgruber, Andreas. 1981. *Germany and the Two World Wars.* Cambridge, MA: Harvard University Press.

Hindley, Meredith. 1996. "Negotiating the Boundary of Unconditional Surrender: The War Refugee Board in Sweden and Nazi Proposals to Ransom Jews, 1944–1945." *Holocaust and Genocide Studies* 10(1): 52–77.

Hinsley, F. H. 1979. *British Intelligence in the Second World War: Its Influence on Strategy and Operations.* London: Her Majesty's Stationery Office.

Hiro, Dilip. 1989. *The Longest War: The Iran-Iraq Military Conflict.* London: Grafton Books.

Hitler, Adolf. [1925] 1999. *Mein Kampf.* Translated by Ralph Manheim. Boston: Houghton Mifflin.

———. [1928] 2003. *Second Book.* New York: Enigma.

Hjeholt, Holger. 1965. "British Mediation in the Danish-German Conflict, 1848–1850: Part One, From the March Revolution to the November Government." *Historisk-filosofisk Meddelelser* 41(1): 5–235.

———. 1966. "British Mediation in the Danish-German Conflict, 1848–1850: Part Two, From the November Cabinet until the Peace with Prussia and the London Protocol (the 2nd of July and the 2nd of August, 1850)." *Historisk-filosofisk Meddelelser* 42(1): 5–251.

Hoffmann, Fritz L., and Ogla Mingo Hoffmann. 1984. *Sovereignty in Dispute: The Falklands/Malvinas, 1493–1982.* Boulder, CO: Westview Press.

Holt, Edgar. 1964. *The Opium Wars in China.* London: Putnam.

Horvath, William J. 1968. "A Statistical Model for the Duration of Wars and Strikes." *Behavioral Science* 13: 18–28.

Houweling, Henk, and Jan G. Siccama. 1988. "Power Transitions as a Cause of War." *Journal of Conflict Resolution* 32(1): 87–102.

Howard, Harry N. 1931. *The Partition of Turkey: A Diplomatic History, 1913–1923.* Norman: University of Oklahoma Press.

Hughes, Wayne Jr., and Jeffrey A. Larson. 1985. *The Falklands Wargame.* Bethesda, MD: US Army Concepts Analysis Agency.

Hull, Cordell. 1948. *The Memoirs of Cordell Hull.* New York: Macmillan.

Humphreys, R. A. 1957. "The Emancipation of Latin America." In *The New Cambridge Modern History, Volume IX: War and Peace in an Age of Upheaval*, edited by J. O. Lindsay, 612–638. Cambridge: Cambridge University Press.

Hunt, G. H., and George Townsend. 1858. *Outram and Havelock's Persian Campaign, To Which Is Prefixed a Summary of Persian History, an Account of Various Differences between England and Persia, and an Inquiry into the Origins of the Late War.* London: G. Routledge.

Huntington, Samuel P. 1993. "The Clash of Civilizations?" *Foreign Affairs* 72(3): 22–49.

———. 1996. *The Clash of Civilizations and the Remaking of World Order.* New York: Simon and Schuster.

Hurd, Douglas. 1967. *The Arrow War: An Anglo-Chinese Confusion, 1856–1860*. London: Collins.

Ike, Nobutaka. 1967. *Japan's Decision for War: Records of the 1941 Policy Conferences*. Stanford: Stanford University Press.

Iklé, Fred Charles. 1991. *Every War Must End*. New York: Columbia University Press.

Ishihara, Shintaro. 1991. *The Japan that Can Say No*. New York: Simon & Schuster.

Jackson, Julian. 2003. *The Fall of France: The Nazi Invasion of 1940*. Oxford: Oxford University Press.

Jackson, Matthew O., and Massimo Morelli. 2009. "The Reasons for Wars: An Updated Survey." In *The Handbook on the Political Economy of War*, edited by Christopher J. Coyne and Rachel L. Mathers, 34–57. Cheltenham, UK: Edward Elgar.

James, Patrick, and John R. Oneal. 1991. "The Influence of Domestic and International Politics on the President's Use of Force." *Journal of Conflict Resolution* 35(2): 307–332.

James, Patrick, and Jean Sébastien Rioux. 1998. "International Crises and Linkage Politics: The Experiences of the United States, 1953–1994." *Political Research Quarterly* 51(3): 781–812.

James, Robert Rhodes. 1970. *Churchill: A Study in Failure, 1900–1939*. London: Weidenfeld and Nicolson.

Jarausch, Konrad H. 1969. "The Illusion of Limited War: Chancellor Bethmann-Hollweg's Calculated Risk, July 1914." *Central European History* 2(1): 48–76.

Jervis, Robert. 1976. *Perception and Misperception in International Politics*. Princeton: Princeton University Press.

——. 1978. "Cooperation under the Security Dilemma." *World Politics* 30(2): 168–214.

——. 1993. "Images and the Gulf War." In *The Political Psychology of the Gulf War: Leaders, Publics, and the Process of Conflict*, edited by Stanley A. Renshon, 173–179. Pittsburgh: University of Pittsburgh Press.

——. 2006. "The Politics and Psychology of Intelligence and Intelligence Reform." *The Forum: A Journal of Applied Research in Contemporary Politics* 4(1): 1–9.

Johnson, Dominic D. P. 2004. *Overconfidence and War: The Havoc and Glory of Positive Illusions*. Cambridge, MA: Harvard University Press.

Kahn, David. 1978. *Hitler's Spies: German Military Intelligence in World War II*. New York: Collier Books.

Kant, Immanuel. [1795] 1957. *Perpetual Peace*. New York: Liberal Arts Press.

Karabell, Zachary. 1995. "Backfire: US Policy towards Iraq, 1988–2 August 1990." *Middle East Journal* 49(1): 28–47.

Karsh, Efraim. 1987–88. "Military Power and Foreign Policy Goals: The Iran-Iraq War Revisited." *International Affairs* 64(1): 83–95.

Kasozi, A.B.K., Nakanyike Musisi, and James Mukooza Sejjengo. 1994. *The Social Origins of Violence in Uganda: 1964–1985*. Montreal: McGill-Queen's University Press.

Katra, William H. 1996. *The Argentine Generation of 1837: Echeverría, Alberdi, Sarmiento, Mitre*. Madison, NJ: Fairleigh Dickinson University Press.

Kaye, John William. 1864. *A History of the Sepoy War in India, 1857–1858*. London: W. H. Allen.

Kecskemeti, Paul. 1958. *Strategic Surrender: The Politics of Victory and Defeat*. Stanford: Stanford University Press.

Keegan, John. 1990. *The Second World War*. New York: Penguin Books.

Kennedy, Paul M. 1987. *The Rise and Fall of the Great Powers, 1500–2000*. New York: Random House.

Kerner, Robert J. 1937. "Russia's New Policy in the Near East after the Peace of Adrianople; Including the Test of the Protocol of 16 September 1829." *Cambridge Historical Journal* 5(3): 280–290.

Kershaw, Ian. 1985. *The Nazi Dictatorship: Problems and Perspectives of Interpretation.* London: Edward Arnold.

——. 1987. *The 'Hitler Myth': Image and Reality in the Third Reich.* Oxford: Oxford University Press.

——. 1998. *Hitler, 1889–1936: Hubris.* London: Allen Lane.

——. 2000. *Hitler, 1936–1945: Nemesis.* New York: W. W. Norton.

Khadduri, Majid. 1988. *The Gulf War: The Origins and Implications of the Iraq-Iran Conflict.* New York: Oxford University Press.

Khadduri, Majid, and Edmund Ghareeb. 1997. *War in the Gulf, 1990–91: The Iraq-Kuwait Conflict and Its Implications.* New York: Oxford University Press.

Khoury, Philip S. 1987. *Syria and the French Mandate: The Politics of Arab Nationalism, 1920–1945.* Princeton: Princeton University Press.

Kim, Woosang, and James D. Morrow. 1992. "When Do Power Shifts Lead to War?" *American Journal of Political Science* 36(4): 896–922.

Kimball, Warren F. 1997. *Forged in War: Roosevelt, Churchill, and the Second World War.* New York: William Morrow.

King, Bolton. 1967. *A History of Italian Unity. Volume 2.* New York: Russell & Russell.

King, Ralph. 1987. *The Iran-Iraq War: The Political Implications.* Adelphi Papers 119. London: The International Institute for Strategic Studies.

Kirshner, Jonathan. 2000. "Rationalist Explanations for War?" *Security Studies* 10(1): 143–150.

Klein, Burton H. 1959. *Germany's Economic Preparations for War.* Cambridge, MA: Harvard University Press.

Kleinpenning, Jan M. G. 2002. "Strong Reservations about 'New Insights into the Demographics of the Paraguayan War.'" *Latin American Research Review* 37(3): 137–142.

Knight, Jonathan. 1977. "Churchill and the Approach to Mussolini and Hitler in May 1940: A Note." *British Journal of International Studies* 3(1): 92–96.

Knights, Michael Andres. 2005. *Cradle of Conflict: Iraq and the Birth of Modern U.S. Military Power.* Annapolis: Naval Institute Press.

Koch, H. W. 1968. "Hitler and the Origins of the Second World War: Second Thoughts on the Origins of Some of the Documents." *Historical Journal* 11(1): 125–143.

——. 1975. "The Specter of a Separate Peace in the East: Russo-German 'Peace Feelers,' 1942–44." *Journal of Contemporary History* 10(3): 531–549.

Kolinski, Charles J. 1965. *Independence or Death! The Story of the Paraguayan War.* Gainesville: University of Florida Press.

Kostiner, Joseph. 1993. "Kuwait: Confusing Friend and Foe." In *Iraq's Road to War*, edited by Amatzia Baram and Barry Rubin, 105–116. New York: St. Martin's Press.

Kraay, Hendrik. 2004. "Patriotic Mobilization in Brazil: The Zuavos and Other Black Companies." In *I Die with My Country: Perspectives on the Paraguayan War, 1864–1870*, edited by Hendrik Kraay and Thomas Whigham. Lincoln: University of Nebraska Press, pp. 61–80.

Krosnick, Jon A. 1990. "Government Policy and Citizen Passion: A Study of Issue Publics in Contemporary America." *Political Behavior* 12(1): 59–92.

Kugler, Jacek, and Douglas Lemke. 1996. *Parity and War: Evaluations and Extensions of the War Ledger.* Ann Arbor: University of Michigan Press.

Kupchan, Charles A. 1994. *The Vulnerability of Empire.* Ithaca: Cornell University Press.

Kydd, Andrew H. 2003. "Which Side Are You On? Bias, Credibility, and Mediation." *American Journal of Political Science* 47(4): 597–611.

——. 2005. *Trust and Mistrust in International Relations*. Princeton: Princeton University Press.

——. 2006. "When Can Mediators Build Trust?" *American Political Science Review* 100(3): 449–462.

Lacina, Bethany, and Nils Petter Gleditsch. 2005. "Monitoring Trends in Global Combat: A New Dataset of Battle Deaths." *European Journal of Population* 21(2–3): 145–166.

Lai, Brian, and Dan Reiter. 2005. "Rally 'Round the Union Jack? Public Opinion and the Use of Force in the United Kingdom, 1948–2001." *International Studies Quarterly* 49(2): 255–272.

Langer, William L., and S. Everett Gleason. 1953. *The Undeclared War, 1940–1941*. New York: Harper & Brothers.

Lawlor, Sheila. 1994. *Churchill and the Politics of War, 1940–1941*. London: Cambridge University Press.

Lawson, Philip. 1993. *The East India Company: A History*. London: Longman.

Lebow, Richard Ned. 1985. "Miscalculation in the South Atlantic: The Origins of the Falkland War." In *Psychology and Deterrence*, edited by Robert Jervis, Richard Ned Lebow, and Janice Gross Stein, 89–124. Baltimore: Johns Hopkins University Press.

Lee, Chong-Sik. 1983. *Revolutionary Struggle in Manchuria: Chinese Communism and Soviet Interest, 1922–1945*. Berkeley: University of California Press.

Leffler, Melvyn P. 1992. *A Preponderance of Power: National Security, the Truman Administration, and the Cold War*. Stanford: Stanford University Press.

Lemke, Douglas. 2002. *Regions of War and Peace*. Cambridge: Cambridge University Press.

Lenin, V. I. 1920. *Imperialism, the Highest Stage of Capitalism*. Moscow: Foreign Languages Publishing House.

Leuchars, Chris. 2002. *To the Bitter End: Paraguay and the War of the Triple Alliance*. Westport, CT: Greenwood Press.

Leventoglu, Bahar, and Branislav L. Slantchev. 2007. "The Armed Peace: A Punctuated Equilibrium Model of War." *American Journal of Political Science* 51(4): 755–771.

Levy, Jack S. 1983. *War in the Modern Great Power System, 1495–1975*. Lexington: University Press of Kentucky.

——. 1985. "Theories of General War." *World Politics* 37(3): 344–374.

——. 1987. "Declining Power and the Preventive Motivation for War." *World Politics* 40(1): 82–107.

——. 1989. "The Diversionary Theory of War: A Critique." In *Handbook of War Studies*, edited by Manus I. Midlarsky, 259–288. Boston: Unwin Hyman.

Levy, Jack S., and Lily I. Vakili. 1992. "Diversionary Action by Authoritarian Regimes: Argentina in the Falklands/Malvinas Case." In *The Internationalization of Communal Strife*, edited by Manus I. Midlarsky, 118–146. London: Routledge.

Lewis, Daniel K. 2001. *The History of Argentina*. Westport, CT: Greenwood Press.

Lian, Bradley, and John R. Oneal. 1993. "Presidents, the Use of Military Force, and Public Opinion." *Journal of Conflict Resolution* 37(2): 277–300.

Licklider, Roy. 1995. "The Consequences of Negotiated Settlements in Civil Wars, 1945–1993." *American Political Science Review* 89(3): 681–690.

Lieber, Kier A. 2007. "The New History of World War I and What It Means for International Relations Theory." *International Security* 32(2): 155–191.

Lincoln, W. Bruce. 1978. *Nicholas I: Emperor and Autocrat of All the Russians*. Bloomington: Indiana University Press.

Lloyd, Alan. 1977. *Destroy Carthage! The Death Throes of an Ancient Culture*. London: Souvenir Press.

Long, Jerry M. 2004. *Saddam's War of Words: Politics, Religion, and the Iraqi Invasion of Kuwait*. Austin: University of Texas Press.

Lüdtke, Alf. 1992. "The Appeal of Exterminating 'Others': German Workers and the Limits of Resistance." *Journal of Modern History* 64 (Supplement: Resistance against the Third Reich): S46–S67.

Lukacs, John. 1999. *Five Days in London, May 1940*. New Haven: Yale University Press.

Lynch, John. 1981. *Argentine Dictator: Juan Manuel De Rosas, 1829–1852*. Oxford: Oxford University Press.

Mack, Andrew. 1975. "Why Big Nations Lose Small Wars: The Politics of Asymmetric Conflict." *World Politics* 27(2): 175–200.

MacLean, David. 1995. *War, Diplomacy, and Informal Empire: Britain and the Republics of La Plata, 1836–1853*. London: British Academic Press.

Maddison, Angus. 2003. *The World Economy: Historical Statistics*. Paris: Organization for Economic Cooperation and Development.

Mambo, Andrew, and Julian Schofield. 2007. "Military Diversion in the 1978 Uganda-Tanzania War." *Journal of Political and Military Sociology* 35(2): 299–321.

Mamdani, Mahmood. 1983. *Imperialism and Fascism in Uganda*. Trenton, NJ: Africa World Press.

Mansfield, Edward D., and Jack L. Snyder. 2005. *Electing to Fight: Why Emerging Democracies Go to War*. Cambridge: MIT Press.

Marashi, Ibrahim al-. 2003. "The Struggle for Iraq: Understanding the Defense Strategy of Saddam Hussein." *Middle East Review of International Affairs* 7(2): 1–10.

Marder, Arthur J. 1981. *Old Friends, New Enemies: The Royal Navy and the Imperial Japanese Navy*. Oxford: Clarendon Press.

Marshall, Monty G., Keith Jaggers, and Ted Robert Gurr. 2010. "Polity IV Project Dataset Users' Manual: Political Regime Characteristics and Transitions, 1800–2010." Vienna, VA: Center for Systemic Peace.

Martin, Vanessa. 2005. *The Qajar Pact: Bargaining, Protest, and the State in Nineteenth-Century Persia*. London: I. B. Tauris.

Mason, T. David, and Patrick J. Fett. 1996. "How Civil Wars End: A Rational Choice Approach." *Journal of Conflict Resolution* 40(4): 546–568.

Mason, Timothy. 1981. "Intention and Explanation: A Current Controversy about the Interpretation of National Socialism." In *The "Führer State," Myth and Reality: Studies on the Structure and Politics of the Third Reich*, edited by Gerhard Hirschfeld and Lothar Kettenacker, 3–20. Stuttgart: Klett-Cota.

Mastny, Vojtech. 1972. "Stalin and the Prospects of a Separate Peace in World War II." *American Historical Review* 77(2): 1365–1388.

Matar, Fuad. 1981. *Saddam Hussein: The Man, the Cause, and the Future*. London: Third World Centre for Research and Publishing.

May, Ernest B. 2000. *Strange Victory: Hitler's Conquest of France*. New York: Hill and Wang.

McCrum, Robert. 1978. "French Rhineland Policy at the Paris Peace Conference, 1919." *Historical Journal* 21(3): 623–648.

McLynn, F. J. 1979. "The Causes of the War of the Triple Alliance: An Interpretation." *Inter-American Economic Affairs* 33(2): 21–43.

Mearsheimer, John J. 1990. "Back to the Future: Instability in Europe after the Cold War." *International Security* 15(4): 5–56.

Michaelis, Meir. 1972. "World Power Status or World Dominion? A Survey of the Literature on Hitler's 'Plan of World Dominion' (1937–1970)." *Historical Journal* 15(2): 331–360.

Miwa, Kimitada. 1975. "Japanese Images of War with the United States." In *Mutual Images: Essays in American-Japanese Relations*, edited by Akira Iriye, 115–137. Cambridge: Harvard University Press.

Molnár, Miklós. 1971. *Budapest 1956: A History of the Hungarian Revolution*. London: George Allen & Unwin.

Moltmann, Günter. 1961. "Weltherrschaftsideen Hitlers." In *Europa und Übersee: Festschrift für Egmont Zechlin*, edited by Otto Brunner and Dietrich Gerhard, 197–240. Hamburg: Verlag Hans Bredow-Institut.

Mommsen, Hans. 1991. *From Weimar to Auschwitz*. Princeton: Princeton University Press.

Montgomery, A. E. 1972. "The Making of the Treaty of Sèvres of 10 August 1920." *Historical Journal* 15(4): 775–787.

Moore, Will H., and David J. Lanoue. 2003. "Domestic Politics and U.S. Foreign Policy: A Study of Cold War Conflict Behavior." *Journal of Politics* 65(2): 376–396.

Morrison, Donald G., and David C. Schmittlein. 1980. "Jobs, Strikes, and Wars: Probability Models for Duration." *Organizational Behavior and Human Performance* 25: 224–251.

Mueller, John E. 1970. "Presidential Popularity from Truman to Johnson." *American Political Science Review* 64(1): 18–34.

——. 1973. *War, Presidents, and Public Opinion*. New York: Wiley.

——. Mueller, John. 1980. "The Search for the Breaking Point in Vietnam: The Statistics of a Deadly Quarrel." *International Studies Quarterly* 24(4): 497–519.

——. 1989. *Retreat from Doomsday: The Obsolescence of Major War*. New York: Basic Books.

——. 1994. *Policy and Opinion in the Gulf War*. Chicago: University of Chicago Press.

——. 2004. *The Remnants of War*. Ithaca: Cornell University Press.

Nakache, Karen. 1999. *La France et le Levant de 1918 à 1923: La Sort de la Cilicie et de ses Confins Militaires*. Ph.D. diss., Université de Nice Sophia-Antipolis.

Needell, Jeffrey D. 2006. *The Party of Order: The Conservatives, the State, and Slavery in the Brazilian Monarchy, 1831–1871*. Stanford: Stanford University Press.

Neillands, Robin. 2001. *The Bomber War: Arthur Harris and the Allied Bomber Offensive, 1939–1945*. London: John Murray.

Neumann, Iver B. 1996. *Russia and the Idea of Europe: A Study in Identity and International Relations*. London: Routledge.

Nincic, Miroslav. 1990. "U.S. Soviet Policy and the Electoral Connection." *World Politics* 42(3): 370–396.

Nott, John. 2005. "A View from the Centre." In *The Falklands Conflict Twenty Years On: Lessons for the Future*, edited by Stephen Badsey, Rob Havers, and Mark Grove, 57–63. London: Frank Cass.

Notter, Harley A. 1949. *Postwar Foreign Policy Preparation, 1939–1945*. Washington, DC: United States Department of State.

Nye, Joseph S. 2004. *Soft Power: The Means to Success in World Politics*. New York: Public Affairs.

Oakes, Amy. 2006. "Diversionary War and Argentina's Invasion of the Falkland Islands." *Security Studies* 15(3): 431–463.

O'Neill, Barry. 2001. "Risk Aversion in International Relations Theory." *International Studies Quarterly* 45(4): 617–640.

Oren, Michael B. 2002. *Six Days of War: June 1967 and the Making of the Modern Middle East*. New York: Oxford University Press.

Organski, A.F.K. 1968. *World Politics*. New York: Alfred A. Knopf.

Organski, A.F.K., and Jacek Kugler. 1980. *The War Ledger*. Chicago: University of Chicago Press.

Ostrom, Charles W. Jr., and Brian L. Job. 1986. "The President and the Political Use of Force." *American Political Science Review* 80(2): 541–566.

Outram, James. 1860. *Lieut.-General Sir James Outram's Persian Campaign in 1857*. London: Smith, Elder.

Overy, Richard. 1999. "Germany and the Munich Crisis: A Mutilated Victory?" In *The Munich Crisis, 1938: Prelude to World War II*, edited by Igor Lukes and Erik Goldstein, 191–215. London: Frank Cass.

———. 2001. *The Battle of Britain: The Myth and the Reality*. New York: W. W. Norton.

Owen IV, John M. 2010. *The Clash of Ideas in World Politics: Transnational Networks, States, and Regime Change, 1510–2010*. Princeton: Princeton University Press.

Pastore, Mario. 1994. "State-Led Industrialization: The Evidence on Paraguay, 1852–1870." *Journal of Latin American Studies* 26(2): 295–324.

Pedroncini, Guy. 1967. *Les Mutineries de 1917*. Paris: Presses Universitaires de France.

Pelletiere, Stephen C. 1992. *The Iran-Iraq War: Chaos in a Vacuum*. New York: Praeger.

Peterson, Harold F. 1932. "Efforts of the United States to Mediate in the Paraguayan War." *Hispanic American Historical Review* 12(1): 2–17.

Phelps, Gilbert. 1975. *Tragedy of Paraguay*. London: Charles Knight.

Pillar, Paul. 1983. *Negotiating Peace: War Termination as a Bargaining Process*. Princeton: Princeton University Press.

Pion-Berlin, David. 1985. "The Fall of Military Rule in Argentina, 1976–1983." *Journal of Interamerican Studies and World Affairs* 27(2): 55–76.

Plá, Josefina. 1976. *The British in Paraguay, 1850–1870*. Richmond, UK: Richmond Publishing.

Pleshakov, Constantine. 2005. *Stalin's Folly: The Tragic First Ten Days of World War II on the Eastern Front*. Boston: Houghton Mifflin.

Powell, Robert. 1999. *In the Shadow of Power: States and Strategies in International Relations*. Princeton: Princeton University Press.

———. 2002. "Bargaining Theory and International Conflict." *Annual Review of Political Science* 5: 1–30.

———. 2004. "Bargaining and Learning While Fighting." *American Journal of Political Science* 48(2): 344–361.

———. 2006. "War as a Commitment Problem." *International Organization* 60(1): 169–203.

Power, Samantha. 2002. *A Problem from Hell: America and the Age of Genocide*. New York: Basic Books.

Pozdeeva, Lydia V. 1994. "The Soviet Union: Territorial Diplomacy." In *Allies at War: The Soviet, American, and British Experience, 1939–1945*, edited by David Reynolds, Warren F. Kimball, and A. O. Chubarian, 355–385. New York: St. Martin's Press.

Press, Daryl G. 2004–5. "The Credibility of Power: Assessing Threats during the 'Appeasement' Crises of the 1930s." *International Security* 29(3): 136–169.

Pulcini, Theodore. 2003. "Russian Orthodoxy and Western Christianity: Confrontation and Accommodation." In *Russia and Western Civilization: Cultural and Historical Encounters*, edited by Russell Bova, 78–112. Armonk, NY: M.E. Sharpe.

Puryear, Vernon John. 1935. *International Economics and Diplomacy in the Near East: A Study of British Commercial Policy in the Levant, 1834–1853*. Stanford: Stanford University Press.

Quinlivan, James T. 1999. "Coup-Proofing: Its Practice and Consequences in the Middle East." *International Security* 24(2): 131–165.

Rauschning, Hermann. 1940. *The Voice of Destruction*. New York: G.P. Putnam's Sons.

Rawlinson, Henry Creswick. 1875. *England and Russia in the East: A Series of Papers on the Political and Geographic Condition of Central Asia*. London: J. Murray.

Reber, Vera Blinn. 1988. "The Demographics of Paraguay: A Reinterpretation of the Great War, 1864–70." *Hispanic American Historical Review* 68(2): 289–319.

———. 2002. "Comment on 'The Paraguayan Rosetta Stone.'" *Latin American Research Review* 37(3): 129–136.

Reed, William. 2003. "Information, Power, and War." *American Political Science Review* 97(4): 633–641.

Reese, Roger. 1989. "A Note on a Consequence of the Expansion of the Red Army on the Eve of World War II." *Soviet Studies* 41(1): 135–140.

Reiter, Dan. 1995. "Exploding the Powder Keg Myth: Preemptive Wars Almost Never Happen." *International Security* 20(2): 5–34.

———. 2003. "Exploring the Bargaining Model of War." *Perspectives on Politics* 1(1): 27–43.

———. 2009. *How Wars End*. Princeton: Princeton University Press.

Reiter, Dan, and Curtis Meek. 1999. "Determinants of Military Strategy, 1904–1999: A Quantitative Empirical Test." *International Studies Quarterly* 43: 363–387.

Reiter, Dan, and Allan C. Stam. 2002. *Democracies at War*. Princeton: Princeton University Press.

Reynolds, David. 1985. "Churchill and the British 'Decision' to Fight on in 1940: Right Policy, Wrong Reasons." In *Diplomacy and Intelligence during the Second World War*, edited by Richard Langhorne, 147–167. Cambridge: Cambridge University Press.

———. 1990. "1940: Fulcrum of the Twentieth Century?" *International Affairs (Royal Institute of International Affairs, 1944–)* 66(2): 325–350.

———. 2001. "Churchill's Writing of History: Appeasement, Autobiography, and 'The Gathering Storm.'" *Transactions of the Royal Historical Society* 6th Ser., 11: 221–247.

———. 2006. *From World War to Cold War: Churchill, Roosevelt, and the International History of the 1940s*. Oxford: Oxford University Press.

Rich, Norman. 1973–74. *Hitler's War Aims: Ideology, the Nazi State, and the Course of Expansion*. New York: W. W. Norton.

———. 1985. *Why the Crimean War? A Cautionary Tale*. New York: McGraw-Hill.

Richards, Diana, T. Clifton Morgan, Rick K. Wilson, Valerie L. Schwebach, and Garry D. Young. 1993. "Good Times, Bad Times, and the Diversionary Use of Force: A Tale of Some Not-So-Free Agents." *Journal of Conflict Resolution* 37(3): 504–535.

Richardson, Lewis F. 1960. *Statistics of Deadly Quarrels*. Pittsburgh: Boxwood Press.

Ripsman, Norrin M., and Jack S. Levy. 2008. "Wishful Thinking or Buying Time? The Logic of British Appeasement in the 1930s." *International Security* 33(2): 148–181.

Ritter, Gerhard. 1958. *The Schlieffen Plan: Critique of a Myth*. London: O. Wolff.

Roberts, Andrew. 1991. *"The Holy Fox": A Biography of Lord Halifax*. London: Weidenfeld and Nicolson.

Roberts, Geoffrey. 2006. *Stalin's Wars: From World War to Cold War, 1939–1953*. New Haven: Yale University Press.

Robins, Philip. 1989. "Iraq in the Gulf War: Objectives, Strategies and Problems." In *The Gulf War: Regional and International Dimensions*, edited by Hanns W. Maull and Otto Pick, 45–59. New York: St. Martin's Press.

Rock, David. 1985. *Argentina, 1516–1982: From Spanish Colonization to the Falklands War*. Berkeley: University of California Press.

Rock, Stephen R. 1989. *Why Peace Breaks Out: Great Power Rapprochement in Historical Perspective*. Chapel Hill: University of North Carolina Press.

Rolland, Denis. 2005. *La Grève Des Tranchées: Les Mutineries de 1917*. Paris: Imago.

Roosevelt, Elliott. 1946. *As He Saw It*. New York: Duell, Sloan and Pearce.

Ross, Michael L. 2004. "How Do Natural Resources Influence Civil War? Evidence from Thirteen Cases." *International Organization* 58(1): 35–67.

Rothfels, Hans. 1961. *The German Opposition to Hitler: An Assessment*. London: O. Wolff.

Rubin, Barry. 1993. "The United States and Iraq: From Appeasement to War." In *Iraq's Road to War*, edited by Amatzia Baram and Barry Rubin, 255–272. New York: St. Martin's Press.

Runciman, Steven. 1951. *A History of the Crusades, Volume I: The First Crusade and the Foundation of the Kingdom of Jerusalem*. Cambridge: Cambridge University Press.

Russett, Bruce M., John R. Oneal, and Michaelene Cox. 2000. "Clash of Civilizations, or Realism and Liberalism Déjà Vu? Some Evidence." *Journal of Peace Research* 37(5): 583–608.

Saakian, R. G. 1986. *Franko-Turetskie Otnosheniia i Kilikiia v 1918–1923 gg*. Erevan: Izdatel'stvo AN Armianskoi SSR.

Saeger, James Schofield. 2007. *Francisco Solano López and the Ruination of Paraguay: Honor and Egocentrism*. New York: Rowman & Littlefield.

Sagan, Scott D. 1988. "The Origins of the Pacific War." In *The Origin and Prevention of Major Wars*, edited by Robert I. Rotberg and Theodore K. Rabb, 323–352. Cambridge: Cambridge University Press.

Sarkees, Meredith Reid, and Frank Wayman. 2010. *Resort to War: 1816–2007*. Washington, DC: CQ Press.

Sarmiento, Domingo F. [1845] 1998. *Facundo: Or, Civilization and Barbarism*. New York: Penguin Books.

Scheina, Robert L. 2003. *Latin America's Wars: The Age of the Caudillo, 1791–1899*. Washington, DC: Brassey's.

Schelling, Thomas C. 1966. *Arms and Influence*. New Haven: Yale University Press.

Schlafley, Phyllis. 1999. "Clinton's Post-Impeachment Push for Power: How Clinton Is Using Kosovo." *Phyllis Schlafly Report* 32(8): 1.

Schroeder, Paul W. 1972. *Austria, Great Britain, and the Crimean War: The Destruction of the European Concert*. Ithaca: Cornell University Press.

——. 1994. *The Transformation of European Politics, 1763–1848*. New York: Clarendon Press.

Schuessler, John M. 2009. "The Deception Dividend? FDR's Undeclared War." *International Security* 34(4): 133–165.

Schulman, Milton. 1948. *Defeat in the West*. New York: E. P. Dutton.

Schwarzkopf, H. Norman and Peter Petre. 1992. *General H. Norman Schwarzkopf: The Autobiography: It Doesn't Take a Hero*. New York: Bantam Books.

Schweller, Randall L. 1994. "Bandwagoning for Profit: Bringing the Revisionist State Back In." *International Security* 19(1): 72–107.

——. 1998. *Deadly Imbalances: Tripolarity and Hitler's Strategy for World Conquest*. New York: Columbia University Press.

——. 2006. *Unanswered Threats: Political Constraints on the Balance of Power*. Princeton: Princeton University Press.

Seaton, Albert. 1977. *The Crimean War: A Russian Chronicle*. New York: St. Martin's Press.

Seifzadeh, Hossein S. 1997. "Revolution, Ideology, and the War." In *Iranian Perspectives on the Iran-Iraq War*, edited by Farhang Rajaee, 90–97. Gainesville: University Press of Florida.

Self, Robert. 2006. *Neville Chamberlain: A Biography*. Aldershot, UK: Ashgate.

Seward, Desmond. 1978. *The Hundred Years War: The English in France, 1337–1453*. New York: Atheneum.

Sherwood, Robert E. 1948. *The White House Papers of Harry L. Hopkins: An Intimate History*. London: Eyre & Spottiswoode.

——. 1950. *Roosevelt and Hopkins: An Intimate History*. New York: Harper.

Shore, Zachary. 2003. *What Hitler Knew: The Battle for Information in Nazi Foreign Policy*. Oxford: Oxford University Press.

Simmel, Georg. 1898. "The Persistence of Social Groups." *American Journal of Sociology* 3(5): 662–698.

——. 1955. *Conflict*. Glencoe, IL: Free Press.

Simpson, John. 1991. *From the House of War: John Simpson in the Gulf*. London: Hutchinson.

Singer, J. David, Stuart Bremer, and John Stuckey. 1972. "Capability Distribution, Uncertainty, and Major Power War, 1820–1965." In *Peace, War, and Numbers*, edited by Bruce Russett, 19–48. Beverly Hills, CA: Sage.

Singer, J. David, and Melvin P. Small. 1972. *The Wages of War, 1816–1965: A Statistical Handbook*. New York: John Wiley.

Slantchev, Branislav L. 2003a. "The Power to Hurt: Costly Conflict with Completely Informed States." *American Political Science Review* 97(1): 123–135.

——. 2003b. "The Principle of Convergence in Wartime Negotiations." *American Political Science Review* 97(4): 621–632.

——. 2004. "How Initiators End Their Wars: The Duration of Conflict and the Terms of Peace." *American Journal of Political Science* 48(4): 813–829.

Slantchev, Branislav L., and Ahmer Tarar. 2011. "Mutual Optimism as a Rationalist Cause of War." *American Journal of Political Science* 55(1): 135–148.

Sloan, Geoffrey. 2005. "The Geopolitics of the Falklands Conflict." In *The Falklands Conflict Twenty Years On: Lessons for the Future*, edited by Stephen Badsey, Rob Havers, and Mark Grove, 23–29. London: Frank Cass.

Sluch, Sergei Zinov'evich. 2004. "Rech' Stalina, Kotoroi Ne Bylo." *Otechestvennaia Istoriia* (1): 113–139.

Smart, Nick. 2003. *British Strategy and Politics during the Phony War: Before the Balloon Went up*. Westport, CT: Praeger.

Smith, Alastair. 1998a. "Fighting Battles, Winning Wars." *Journal of Conflict Resolution* 42(3): 301–320.

——. 1998b. "International Crises and Domestic Politics." *American Political Science Review* 92(3): 623–638.

Smith, Alastair, and Allan C. Stam. 2004. "Bargaining and the Nature of War." *Journal of Conflict Resolution* 48(6): 783–813.

Smith, Hugh. 2005. "What Costs Will Democracies Bear? A Review of Popular Theories of Casualty Aversion." *Armed Forces and Society* 31(4): 487–512.

Smith, Woodruff D. 1986. *The Ideological Origins of Nazi Imperialism*. New York: Oxford University Press.

Snyder, Glenn H., and Paul Diesing. 1977. *Conflict among Nations: Bargaining, Decision Making, and System Structure in International Crisis*. Princeton: Princeton University Press.

Snyder, Jack L. 1991. *Myths of Empire: Domestic Politics and International Ambition*. Ithaca: Cornell University Press.

Sonyel, Salahi Ramsdan. 1975. *Turkish Diplomacy, 1918–1923: Mustafa Kemal and the Turkish Nationalist Movement*. London: Sage.

Sprecher, Christopher, and Karl DeRouen Jr. 2002. "Israeli Military Actions and Internalization-Externalization Processes." *Journal of Conflict Resolution* 46(2): 244–259.

Springhall, John. 2001. *Decolonization since 1945: The Collapse of European Overseas Empires*. New York: Palgrave.

Stafford, David. 1980. *Britain and European Resistance, 1940–1945: A Survey of the Special Operations Executive, with Documents*. London: Macmillan.

———. 2006. "Churchill and SOE." In *Special Operations Executive: A New Instrument of War*, edited by Mark Seaman, 47–60. London: Routledge.

Stanley, Elizabeth A. 2009. *Paths to Peace: Domestic Coalition Shifts, War Termination and the Korean War*. Palo Alto, CA: Stanford University Press.

Stanley, Elizabeth A., and John P. Sawyer. 2009. "The Equifinality of War Termination: Multiple Paths to Ending War." *Journal of Conflict Resolution* 53(5): 651–676.

Stedman, Stephen John. 1997. "Spoiler Problems in Peace Processes." *International Security* 22(2): 5–53.

Stein, Janice Gross. 1992. "Deterrence and Compellence in the Gulf, 1990–1991: A Failed or Impossible Task?" *International Security* 17(2): 147–179.

Stewart, Watt, and Harold F. Peterson. 1942. *Builders of Latin America*. New York: Harper & Brothers.

Stiglitz, Joseph E., and Linda J. Bilmes. 2008. *The Three Trillion Dollar War: The True Cost of the Iraq Conflict*. New York: W. W. Norton.

Stinnett, Douglas M., Jaroslav Tir, Philip Schafer, Paul F. Diehl, and Charles Gochman. 2002. "The Correlates of War Project Direct Contiguity Data, Version 3." *Conflict Management and Peace Science* 19(2): 58–66.

Tachjian, Vahé. 2004. *La France en Cilicie et en Haute-Mésopotamie: Aux Confins de la Turquie, de la Syrie et de l'Irak (1919–1933)*. Paris: Éditions Karthala.

Tanenbaum, Jan Karl. 1978. "France and the Arab Middle East, 1914–1920." *Transactions of the American Philosophical Society* 68(7): 1–50.

Tarar, Ahmer. 2006. "Diversionary Incentives and the Bargaining Approach to War." *International Studies Quarterly* 50(1): 169–188.

Taylor, A. J. P. 1961. *The Origins of the Second World War*. New York: Simon & Schuster.

———. 1964. *The Habsburg Monarchy, 1809–1918*. New York: Penguin.

Taylor, Telford. 1952. *Sword and Swastika: Generals and Nazis in the Third Reich*. New York: Simon and Schuster.

Thatcher, Margaret. 1993. *The Downing Street Years*. New York: HarperCollins.

Thompson, George. 1869. *The War in Paraguay, with a Historical Sketch of the Country and Its People and Notes upon the Military Engineering of the War*. London: Longmans, Green.

Thucydides. [ca. 400 BCE] 1954. *The Peloponnesian War*. New York: Penguin Classics.

Toase, Francis. 2005. "The United Nations Security Resolution 502." In *The Falklands Conflict Twenty Years On: Lessons for the Future*, edited by Stephen Badsey, Rob Havers, and Mark Grove, 147–169. London: Frank Cass.

Toft, Monica Duffy. 2006. "Issue Indivisibility and Time Horizons as Rationalist Explanations for War." *Security Studies* 15(1): 34–69.

Toynbee, Arnold J. 1922. *The Western Question in Greece and Turkey: A Study in the Contact of Civilizations*. London: Constable.

Trachtenberg, Marc. 1991. *History and Strategy*. Princeton: Princeton University Press.

———. 2005. "The Bush Strategy in Historical Perspective." In *Nuclear Transformation: The New U.S. Nuclear Doctrine*, edited by James Wirtz and Jeffrey Larsen, 9–22. New York: Palgrave Macmillan.

Trask, David F. 1981. *The War with Spain in 1898.* New York: Macmillan.

Trevor-Roper, Hugh Redwald. 1960. "Hitlers Kriegsziele." *Vierteljahrshefte für Zeitgeschichte* 8(2): 121–133.

Tusicisny, Andrej. 2004. "Civilizational Conflicts: More Frequent, Longer, Bloodier?" *Journal of Peace Research* 41(4): 485–498.

Vacs, Aldo C. 1987. "Authoritarian Breakdown and Democratization in Argentina." In *Authoritarians and Democrats: Regime Transition in Latin America,* edited by James M. Malloy and Mitchell A. Seligson, 15–42. Pittsburgh: University of Pittsburgh Press.

Van Dyke, Carl. 1997. *The Soviet Invasion of Finland, 1939–40.* London: Frank Cass.

Van Evera, Stephen. 1998. "Offense, Defense, and the Causes of War." *International Security* 22(4): 5–43.

Villa, Brian L. 1976. "The U.S. Army, Unconditional Surrender, and the Potsdam Proclamation." *Journal of American History* 63(1): 66–92.

Viorst, Milton. 1991. "Report from Baghdad." *New Yorker* 24: 55–73.

von Clausewitz, Carl. 1976. *On War.* Princeton: Princeton University Press.

Vuchinich, Samuel, and Jay Teachman. 1993. "The Duration of Wars, Strikes, Riots, and Family Arguments." *Journal of Conflict Resolution* 37: 544–568.

Wagner, R. Harrison. 2000. "Bargaining and War." *American Journal of Political Science* 44(3): 469–484.

Waller, John H. 1990. *Beyond the Khyber Pass: The Road to British Disaster in the First Afghan War.* New York: Random House.

Walpole, Spencer. 1912. *A History of England from the Conclusion of the Great War in 1815.* London: Longmans, Green.

Walter, Barbara F. 1997. "The Critical Barrier to Civil War Settlement." *International Organization* 51(3): 335–364.

——. 2002. *Committing to Peace: The Successful Settlement of Civil Wars.* Princeton: Princeton University Press.

Waltz, Kenneth N. 1979. *Theory of International Politics.* New York: McGraw-Hill.

Wang, Kevin H. 1996. "Presidential Responses to Foreign Policy Crises: Rational Choice and Domestic Politics." *Journal of Conflict Resolution* 40(1): 68–97.

Warlimont, Walter. 1964. *Inside Hitler's Headquarters, 1939–45.* New York: Praeger.

Warren, Harris Gaylord. 1978. *Paraguay and the Triple Alliance: The Postwar Decade, 1869–1878.* Austin: Institute of Latin American Studies, University of Texas at Austin.

Washburn, Charles A. 1871. *The History of Paraguay, with Notes of Personal Observations, and Reminiscences of Diplomacy under Difficulties.* Boston: Lee and Shepard.

Wawro, Geoffrey. 1996. *The Austro-Prussian War: Austria's War with Prussia and Italy in 1866.* Cambridge: Cambridge University Press.

Wei, Henry. 1956. *China and Soviet Russia.* Princeton: D. Van Nostrand.

Weinberg, Gerhard L. 1964. "Hitler's Image of the United States." *American Historical Review* 69(4): 1006–1021.

——. 1995. *Germany, Hitler, and World War II: Essays in Modern German and World History.* Cambridge: Cambridge University Press.

——. 2005. *Visions of Victory: The Hopes of Eight World War II Leaders.* Cambridge: Cambridge University Press.

Weisiger, Alex. 2012. "Victory without Peace: Conquest, Insurgency, and War Termination." Unpublished manuscript, University of Pennsylvania.

Weiss, Herbert K. 1963. "Stochastic Models for the Duration and Magnitude of a 'Deadly Quarrel'." *Operations Research* 11(1): 101–121.

Wetzel, David. 1985. *The Crimean War: A Diplomatic History.* Boulder, CO: East European Monographs.

Whigham, Thomas. 1991. *The Politics of River Trade: Tradition and Development in the Upper Plata, 1780–1870.* Albuquerque: University of New Mexico Press.

———. 2002. *The Paraguayan War, Volume 1.* Lincoln: University of Nebraska Press.

———. 2004. "The Paraguayan War: A Catalyst for Nationalism in South America." In *I Die with My Country: Perspectives on the Paraguayan War, 1864–1870,* edited by Hendrik Kraay and Thomas Whigham, 179–198. Lincoln: University of Nebraska Press.

Whigham, Thomas L., and Barbara Potthast. 1990. "Some Strong Reservations: A Critique of Vera Blinn Reber's 'The Demographics of Paraguay: A Reinterpretation of the Great War, 1864–70.'" *Hispanic American Historical Review* 70(4): 667–675.

———. 1999. "The Paraguayan Rosetta Stone: New Insights into the Demographics of the Paraguayan War, 1864–1870." *Latin American Research Review* 34(1): 174–186.

Williams, John Hoyt. 1979. *The Rise and Fall of the Paraguayan Republic, 1800–1870.* Austin: Institute of Latin American Studies, University of Texas at Austin.

Wittman, Donald. 1979. "How a War Ends: A Rational Choice Approach." *Journal of Conflict Resolution* 23(4): 743–763.

Wittmer, Felix. 1953. *The Yalta Betrayal: Data on the Decline and Fall of Franklin Delano Roosevelt.* Caldwell, ID: Caxton Printers.

Woods, Kevin M., Michael R. Pease, Mark E. Stout, Williamson Murray, and James G. Lacey. 2006. *The Iraqi Perspective Project: A View of Operation Iraqi Freedom from Saddam's Senior Leadership.* Norfolk, VA: United States Joint Forces Command, Joint Center for Operational Analysis.

Woodward, Bob. 1991. *The Commanders.* New York: Simon & Schuster.

———. 2004. *Plan of Attack.* New York: Simon & Schuster.

Woodward, Sir Llewellyn. 1962. *British Foreign Policy in the Second World War.* London: Her Majesty's Stationary Office.

———. 1971. *British Foreign Policy in the Second World War, Volume II.* London: Her Majesty's Stationary Office.

Wrede-Braden, Antonia Catharina. 2007. *Reputation for Retreat: Casualty-Aversion and the Declining Credibility of Democracies' Threats.* Ph.D. diss., Harvard University.

Wright, Quincy. 1970. "How Hostilities Have Ended: Peace Treaties and Alternatives." *Annals of the American Academy of Political and Social Science* 392: 51–61.

Yagame, Kazuo. 2006. *Konoe Fumimaro and the Failure of Peace in Japan, 1937–1941: A Critical Appraisal of the Three-Time Prime Minister.* Jefferson, NC: McFarland.

Yetiv, Steve A. 2004. *Explaining Foreign Policy: U.S. Decision-Making & the Persian Gulf War.* Baltimore: Johns Hopkins University Press.

Zeidner, Robert F. 2005. *The Tricolor over the Taurus: The French in Cilicia and Vicinity, 1918–1922.* Ankara: Atatürk Supreme Council for Culture, Language and History.

Zürcher, Erik Jan. 1984. *The Unionist Factor: The Rôle of the Committee of Union and Progress in the Turkish National Movement, 1905–1926.* Leiden: E. J. Brill.

Index

[281]

diversionary motive for Argentine junta, 50, 179, 181, 190
  government weakness, 181–83
  invasion as gamble, 183, 189–90
  unsuccessful attempt to prolong war, 188–89
  history of dispute, 179–81
  importance of informational mechanism, 183–86
  limits of preventive war mechanism, 186–87
  quantitative analysis and, 67–68
  research strategy and, 56
  unconditional surrender mechanism, 29
Fieldhouse, John, 186
Fine, Jason P., 75
Finland, 137–38
First Afghan War, 172
First-strike advantage, 25
Flores, Venancio, 88, 90, 96, 97
France
  Allied unconditional surrender strategy and, 122–23
  British pessimism after fall of, 126–28, 244n16
  see also Crimean War; Franco-Turkish War; World War II, in Europe
Francia, José Gaspar de, 95
Franco-Austrian War, 6, 38, 226n70
Franco-Turkish War, 56, 178–79, 190–202, 206–7
  history of, 191, 192f, 193–94, 256nn53, 54
  limits of information mechanism, 200
  limits of preventive war mechanism, 199–200
  principle-agent mechanism, 190, 200–202
  colonial faction in French politics, 194–96, 257n69
  domestic constraints eventually force withdrawal, 197–98
  suppression of information about war, 196–97, 198, 207

Galtieri, Leopoldo, 182, 185, 188–89
George-Picot, François, 195
Germany. *See* World War I; World War II, in Europe
Glaspie, April, 162–63, 251n12
Globalists, German war aims and, 120, 242n79
Godfrey of Bouillon, 222n25
Goebbels, Joseph, 110
Goemans, Hein F., 45, 47, 72

Gorchakov, Alexander, 146
Göring, Hermann, 117–18
Gouraud, Henri, 197, 198, 201
Gray, Robert J., 75
Guatemala, 236n24

Haig, Alexander, 180, 185, 187
Halifax, Lord, 123, 127, 129, 130, 131, 134–35, 243n100, 246n156
Hernández, Fidel Sánchez, 44
Hess, Rudolf, 241n69
Hildebrand, Klaus, 108
Hillgruber, Andreas, 108
Himmler, Heinrich, 118, 125, 242n71
Hirohito, emperor, 46, 152
Hitler, Adolf, 18, 33, 48
  advocacy for preventive war, 113–14
  ideology, 9, 105, 108, 112–14, 133, 139, 208, 239n24, 243n90
  informational mechanism and, 108–10
  invasion of Soviet Union not mistake, 108–19
  Mein Kampf, 112, 116, 119, 120, 121, 122, 133, 238n7, 239n25, 240n51, 243n87, 245n151
  principle-agent mechanism and, 110–11
  racial beliefs, 108, 113, 119–20, 133, 239n24, 240n35, 243n90
  reasons for continuing to fight after Stalingrad, 107, 110, 129, 131
  second, unpublished book, 112, 120, 121, 237n7, 239n25
  unconditional surrender mechanism and, 27, 209–10, 222n38
Hull, Cordell, 130, 132
Hungary, 29, 138, 222n25, 247n47
Huntington, Samuel P., 65, 232n33
Hussein, Saddam
  Iran-Iraq War and, 153–58, 160, 208, 249n47
  Persian Gulf War and, 160–70, 205
  U.S. invasion of Iraq and, 18, 35–36

India, 170, 172, 173, 185, 195, 228n109
Informational mechanism
  conflict management implications, 204–5, 212–13
  defined, 4–5, 7
  divergent expectations, 34–36, 53, 225nn57, 67
  external actors and, 35–36, 37
  misperceptions about resolve, 36, 37

Philippines, U.S. seizure of, 25
Policy wars (nondiversionary principal
  agent conflicts), 7, 43, 201
  quantitative analysis and, 228n1
  see also Franco-Turkish War; Principal-
    agent mechanism
Positive utility, as explanation for war, 15
Powell, Robert, 39, 220nn6, 11
Power transition theory of war, 16
Preventive war mechanism, 11–12, 16–25,
  204, 207–10
  conflict management implications, 213
  defined, 4, 7
  as explanation for unlimited war, 16–25
    duration and intensity, 49f, 52–53
    fighting to conquest, 24
    high war aims, 21, 22–23
    possibility of short but decisive wars,
      20–22, 24
    risky military strategy, 23–24
  as explanation for war, 16–25
    diplomacy as potential alternative to
      war, 18–19, 222n25
    incentives to engage in war, 24–25
    reasons for failure of rising power to
      allay opponent's fears, 18–19
    reasons to fear decline, 3–4, 16–18,
      61–68, 229n15, 230nn17–20
  hypotheses, 22–24
  measuring anticipated decline, 61–63
  quantitative analysis and, 70–72, 75–83,
    207–8
  unconditional surrender and, 25–26,
    30–32
  see also specific case studies
Principal-agent mechanism, 3, 11, 12, 53,
  204, 205–7
  conflict management implications, 213
  defined, 7
  diversionary wars, 43, 46–50, 49f, 178,
    205, 227nn95, 98, 105
  as explanation for limited war, 12,
    42–49
    constraints on leaders, 42–43, 45–49,
      52, 53
  hypotheses, 49–50, 52–53
  internal constraints and, 31, 223n49
  policy wars, 43, 49–52, 49f
  quantitative analysis, 46–47, 58, 72, 78,
    84
  unifies argument about domestic
    politics and war, 43–44
  war duration and intensity, 49–50, 49f
  see also specific case studies

Private information. *See* Informational
  mechanism

Quantitative analysis, 57–85
  central findings, 57–58
  data and measurement, 58–68, 229n15,
    230nn17–20
  statistical tests, 68–83
    disaggregating types of war
      termination, 74–80, 77t, 78f, 79f
    general predictions, 69–74, 69t, 71f,
      73f, 73t
    tests of ancillary hypothesis, 80–83,
      81t

Rauschning, Hermann, 120, 242n80
Realist focus on systemic wars, 6, 219n8
Regime change, unconditional surrender
  and, 29
Reiter, Dan, 27, 82, 222n28, 234n56,
  245n150
Research strategy, 54–57, 228nn1, 2. *See
  also* Quantitative analysis
Reynolds, David, 126
Ribbentrop, Joachim von, 115
Rich, Norman, 147
Risk-acceptance, as explanation for war,
  15
Romania, 138
Roosevelt, Franklin D., 30
  death of, 242n75
  domestic constraints on, 5, 227n104
  unconditional surrender and, 28, 29,
    124, 125, 126, 127, 129–30, 131–32,
    136–37, 137, 138, 151, 243n102
Rosas, Juan Mauel de, 93, 101, 103
Russia. *See* Crimean War; Soviet Union;
  World War II, in Europe
Russo-Finnish War, 109, 115
Russo-Hungarian conflict, 29
Russo-Japanese War, 6, 113
Russo-Polish War, 41
Russo-Turkish War, 143

Saudi Arabia, Persian Gulf War and,
  161–62, 164, 165, 168, 170, 251nn17, 18
Schleswig-Holstein Wars, 41
Schlieffen Plan, 23–24
Second Opium War (Arrow War), 176,
  254n57
Security dilemma, unconditional
  surrender mechanism, 32–33, 224n53
Settlement. *See* Conquest and settlement
Seymour, Lord, 144